GARDNER'S

Film, Video & TV Dictionary

Edited by Garth Gardner, Ph.D.

GARTH GARDNER COMPANY

GGC publishing

Washington DC, USA · London, UK

Team: Ali Lam, Gayle Green, Chris Gosnell, Amiee Beck, Ryan Bell, Malek Z. Abdo.
Art Director: Nic Banks
Cover Designer: Tomoko Miki

Editorial inquiries concerning this book should be e-mailed to: info@ggcinc.com. Website: www.gogardner.com

Library of Congress Cataloging-in-Publication Data

Printed in Canada

Gardner's film, video & TV dictionary / Garth Gardner Company, Inc.
p. cm.
ISBN 1-58965-006-9 (pbk.)
1. Cinematography—Dictionaries. 2. Video recording—Dictionaries.
3. Television—Dictionaries. I. Title: Gardner's film, video, and TV dictionary. II. Garth Gardner Company.
TR847.G37 2002
791.4'03—dc21
2002152070

About this Book

Gardner's Film, Video & TV Dictionary is an essential guide for beginning filmmakers, producers, students, and veterans alike. The book gives concise definitions of the most popular production terms and jargons used in the industry as well as some of the lesser known terms. Entries relating to broadcast, motion picture, and video production form the boundaries of this book making it a must-have reference for anyone in the field. Aesthetic techniques, professional titles, production styles, and other aspects of film, video, and TV are also defined. This reference belongs on the bookshelf of anyone working in the world of film, video or television.

About the Editor

Garth Gardner, Ph.D. is a fine artist and animation producer. He has taught and lectured at several universities including The Ohio State University, William Paterson University, University of California Los Angeles, Fashion Institute of Technology, Florida A&M University, University of Southern California, University of Texas at Austin, Xavier University, and George Mason University. He is a graduate of San Francisco State University and The Ohio State University.

Numbers

1080I: 1080 line, 1920 pixel, widescreen 16: 9 aspect ratio, 30 frames per second format which is interlace scanned.

1080P: 1080 line, 1920 pixel, widescreen 16: 9 aspect ratio, 30 frames per second format which is progressively scanned.

16: 9: The aspect ratio of wide-screen-format television.

180 Rule: This is the rule which states that if two people are filmed in a sequence there is an invisible line between them and the camera should only be positioned anywhere within the 180 degrees on one side of the line. Crossing the line results in a certain particular jump, where is appears that the two people suddenly switched places.

1940P: 1940 line, 4096 pixel, widescreen 16: 9 aspect ratio, 30 frames per second format which is interlace scanned.

24-bit Color: Color for which each red, green, and blue component stores 8 bits of information. 24-bit color is capable of representing over one million different variations of color.

3,200K: 3,200K is the color temperature of Tungsten.

3/4U: Aging industrial videocassette format. Uses 3/4-inch tape in cassettes.

3/4U-SP: Superior performance version of 3/4U, totally compatible with it.

3: 2 Pull-down: The converting of 24 fps film to 30 fps video wherein every other frame of film is held for 3 fields of video. The resulting video has a repeating sequence of 3 fields followed by 2 fields.

300 Ohm: Refers to a usually flat TV antenna cable consisting of two conductors held parallel about 1/2 inch apart by insulating material.

3-D Modeling: Electronic graphics technique or software that allows one to designate points in three dimensional space, connect those points, cover the resulting wire frame with a selected material, then move or rotate the object, showing it from various angles. Objects can be combined and allowed to reflect or cast shadows upon each other and/or their backgrounds.

3-to-1 Rule: A term used to describe position microphones at least 3 times farther from each other than they are from the people speaking into them.

4 FSC: Composite Digital video as used in D2 and D3 VTRs. Stands for 4 times the Frequency of Subcarrier, which is the sampling rate used.

4: 1: 1: The sampling ratio used in the DV or DVC (Digital Video, Digital Videotape) digital video signal. For every 4 samples of luminance there is 1 sample each of the color difference signals, R-Y (Red minus Luminance) and B-Y (Blue minus luminance).

4: 2: 0: The sampling ratio used in some variations of DV digital video signal. For every 4 samples of luminance there is a sample of one of the color difference signals, followed by a sample of the other color difference signal on the next luminance sample.

4: 2: 2: A commonly used term for a component digital video format. The first number is the reference count of luminance pixels. The second number is the corresponding number of color pixels in the odd rows or scan lines. The third number is the corresponding number of color pixels in the even rows. DVD's rating is 4**:** 2**:** 0, the color resolution is half the luminance (advertised; published) resolution both horizontally and vertically, or each two by two block of luminance pixels has to be the same color. For analog video luminance and color resolution vary independently and the color resolution may vary depending on which colors are juxtaposed, so this method cannot be easily applied.

4: 4: 4: A sampling ratio that has equal amounts of the luminance and both chrominance channels.

4:3: The aspect ratio of conventional video, television, and computer screens.

480P: 480 line, 704 pixel, widescreen 16: 9 or 4:3 aspect ratio format 3 which are progressively scanned. Frame rates are 24, 30 or 60 frames per second.

5,400K: The color temperature of daylight.

5.1 Channel Digital Sound: The film digital sound exhibition standard which utilizes five output speaker channels (left, center, right, right surround, left surround, and subwoofer).

56 Kilobit Leased Line: A 56 kilobit (56,000 bits per second) leased line is currently the smallest bandwidth transmission data circuit useful in Internet applications. It is also roughly the bandwidth needed for a voice phone call and will be the new speed of high end modems.

5C: See Digital Transmission Content Protection.

720P: 720 line, 1 280 pixel, widescreen 16: 9 aspect ratio format which is progressively scanned. Frame rates are 24, 30 or 60 frames per second.

75-Ohm: Refers to TV antenna connections made using a round (coaxial) cable. Also refers to the most common video cabling using RCA plugs and jacks. For RCA video cables, almost any cable will transmit a decent picture but critical viewers will have to verify whether the circuits are indeed 75 ohm and purchase cables that match and have adequate bandwidth.

8 mm: A compact videocassette record/playback tape format which uses eight millimeter wide magnetic tape. A worldwide standard established in 1983 allowing high quality video and audio recording. Flexibility, lightweight cameras and reduced tape storage requirements are among the format's advantages.

8-bit Color: A picture's colors are selected from a palette of 256 colors in a color lookup table. Only 256 colors are available in a single picture.

8-pin: Rectangular plug with eight pins and a cable that goes with it used to carry audio, video, and other signals between a monitor/receiver and a videocassette recorder.

8-VSB: Eight discrete amplitude level, (vestigial side-band) broadcast transmission technology. VSB is an analog modulation technique used to reduce the amount of spectrum needed to transmit information through cable TV, or over-the-air broadcasts used in the NTSC (analog) standard. 8-VSB is the U.S. ATSC digital television transmission standard. 8-VSB and COFDM are two possible methods of delivering DTV broadcasts to the home. Each method has advantages which vary depending on the applications for which they are used. 8-VSB is a single carrier modulation system and COFDM is a multi-carrier system. Multiple carriers allow information to be divided among the carriers which, under certain circumstances, makes the system more resistant to multi-path (ghosting) reception conditions.

A

A- and B-rolls: Two reels of source footage, the A roll and the B roll, are used to provide material for editing. This can be used to reduce spooling time and to provide the two simultaneous sources needed for video dissolves, wipes, digital effects etc. The B-roll may be a part or a total copy of the A-roll. This method does not apply in true random access or nonlinear editing where only one copy of the original footage is required.

A/A (A/X/A) Roll Editing: Editing from a single source using effects to transition from the source to itself (source "A" to "A") using a picture freeze at the end of one scene to transition the start of the next scene.

A/B Story: "A" story is the main story/theme while "B" story refers to the background story.

A/B Switch: Electrical switch which selects either the signal from cable A or the signal from cable B and feeds the results to a TV, VCR, or other destination.

A/D: Analog to Digital, as in the audio A/D converter found in the Plum breakout box.

A/V Drive: Audio/video drive; a high-end hard drive capable of storing high-bandwidth (i.e., high data rate) audio/video data.

AAC (Advanced Audio Coder): An audio-encoding standard for MPEG-2 that is not backward-compatible with MPEG-1 audio.

Abby Singer: The second-to-last shot of the day. Named after production manager Abby Singer, who would frequently call "last shot of the day," only to have the director ask for one more take. See also Martini Shot.

Aberration: Any optical or electronic distortion of an image.

Above-the-line Costs: Production expenses related only to a particular show. Examples: special talent, writers, travel, charges for special effects.

Abstract Animation: An aesthetic tenet of animation which sees animation as consisting of lines, shapes and colors, abstract forms to be manipulated by the animator at will. The opposite of naturalistic animation.

AC Adaptor: Device that connects to a wall outlet (AC), and sends power to a device to: charge its batteries, or operate without using battery power.

AC Coupled: AC coupling passes a signal through a capacitor to remove any DC offset, or the overall voltage level that the video signal "rides" on. One way to find the signal is to remove the DC offset by AC coupling, and then do DC restoration to add a known DC offset (one that we selected).

AC: Alternating Current, which comes from the wall outlet. Not DC, direct current, which comes from a battery.

AC'97, AC'98: These are definitions by Intel for the audio I/O implementation for PCs. Two chips are defined: an analog audio I/O chip and a digital controller chip. The digital chip will eventually be replaced by a software solution. The goal is to increase the audio performance of PCs and lower cost.

AC-3: The former name of the Dolby Digital audio-coding system, which is still technically referred to as AC-3 in standards documents.

Academy Aperture: In 35mm this is the full frame exposed by the camera, with an aspect ratio of 1.33. When the film is projected there is a mask in the projector's gate to change the aspect ratio to 1.85 or 1.66, cropping the top and bottom of the image. Older films were not shot to be masked and should be projected without a mask. The Academy Aperture is sometimes called the Full Academy Aperture.

Academy Leader: This is standard countdown leader,

counting down 8 to 3 and then with one frame of 2, at which point there is a single frame beep on the soundtrack. It is used at the beginning of a film for the lab to line up sound (using the beep) and later for the projectionist to know when to turn on the lamp and hopefully not miss the opening of the film. A common mistake is to count the footage from the 2, but actually frame zero is the one right before the first 8, a single frame with the words "Picture Start." Academy Leader is sometimes also known as SMPTE leader.

Academy of Motion Picture Arts and Sciences Awards: The coveted annual award, the "Oscar," which is presented to the best of the best.

Academy of Motion Picture Sound: A UK-based organization whose aims are to promote and encourage the science, technology and creative application of all aspects of motion picture sound recording and reproduction, and to promote and enhance the status and recognition of the contribution of those therein engaged.

Academy Ratio: Academy of Motion Picture Arts and Sciences' standardize shape of the film frame. This standard for 35 mm film is 1.33:1 and means that the height of the frame is 1.33 times the width (0.864 inches x 0.63 inches per frame); the reason for this is there needs to be room on one side (the left) of each frame for the soundtrack. Contemporary widescreen (also known as American widescreen) ratio is 1.85:1 and the Anamorphic widescreen (Panavision) ratio is 2.35:1.

ACB: See Vestigial Sideband.

Acceleration Editing: Condensing or expanding time through editing. In a dramatic production several hours of activity can be condensed into a few short scenes. Conversely, some directors will "drag out" an event beyond its actual time in order to heighten dramatic impact.

Accent Light: A light unit that emphasizes one subject. This might be a key light, a kicker, or a backlight.

Access Channel: Cable TV channel set aside for local community use, like town meetings, school sports, local affairs, and news.

Access Time: The length of time it takes to find, retrieve, and display digital information.

Access: In the context of newsgathering, the legal right a news reporter to go onto private property to obtain information or videotape footage.

Accession Number: Numerical order (1, 2, 3, etc.) assigned to tapes as they are acquired or recorded.

Accessory Mount: Threaded hole on top of camera or camcorder for attaching a light, microphone, or other accessory.

Accrual Balancing: When a television station has billing periods that are different from its accounting periods, and income is reported according to how it was aired, they must use accrual balancing to obtain the correct sales figures for their accounting period. To figure the accrued sales, one must calculate the difference between the total billed for the period and the value of spots that have been aired during the accounting period but have not yet been billed. By adding the accrued sales figure for the current period to the billing totals, and subtracting the previous month's accrued sales figure, one arrives at a sales (income) figure for the accounting period. This figure should balance to a projection for the same period.

ACE: See American Cinematic Editors.

ACES and Deuces: These terms refer to the power of light. ACES are 1K or 1000 watt lights and deuces are 2K or 2000 watt lights.

Acetone: An organic liquid chemical used in the manufacture of some types of film cement. Also used to clean film splicing equipment.

Achromatic: Ability of a high quality lens to not make colored ridges on contrasty objects in the edges of the picture.

Acoustics: The science of the transmission of sound waves. Generally refers to the characteristics of auditoriums, theatres and studios with respect to their design.

Acquisition Format: Typically, footage shot on S-VHS or Hi8 intended to be transferred to a higher quality format in order to retain quality during subsequent editing and copying.

Acrylic Sheet Filters: Filters used outside windows to correct for color temperature of daylight to tungsten light or tungsten to daylight. They have very little reflection and they are optically sound.

Act: A portion or segment of the narrative presented between commercial breaks—consists of one or more scenes.

ACT: See Action for Children's Television.

Action Cutting: The cutting of film from one shot to another in a way yielding the impression that action is continuous and uninterrupted.

Action for Children's Television (ACT): National organization formed by citizens devoted to upgrading the quality of programming and advertising directed toward children.

Action Still: A still-frame captured during a production by an electronic still-store device. Sometimes used at the end of a production to key credits over.

Action: "Action" is called during filming to indicate the start of the current take. See also Cut, Speed, Lock it Down.

Active Lines: The total number of visible scanning lines in a video frame, or the total number of lines, minus the lines devoted to vertical blanking.

Active Matrix: A type of liquid crystal that changes quickly, appropriate for LCD panels that also display video.

Active Picture Area: The part of a TV picture that contains actual image information as opposed to sync or other data. Vertically the active picture area is 486 lines for NTSC and 576 lines for PAL. The inactive area is called Blanking.

Active Video: The part of the video waveform that contains picture information. Most of the active video, if not all of it, is visible on the display.

Active: Electrical device which requires electric power to operate. TV antenna preamplifiers and amplified TV couplers are active.

ActiveMovie: The former name for Microsoft's DirectShow technology.

Actor Movement: Typically referred to by the theatrical term, Blocking.

Actor Star: An actor who can play a wide range of roles. For example, Robert De Niro, Anthony Hopkins, and Emma Thompson are actor stars.

Actor/Actress: The "talent" who plays the role of a character. Historically, the term "Actor" referred exclusively to males, but in modern times the term is used for both genders.

Actual Malice: Broadcasting or publishing something with is either known to be false or with reckless disregard for its truth or accuracy.

Actualities: Events from the historical world used in news and sports programs.

Acutance: The measure of sharpness of a clearly defined object in a scene as reproduced by a lens.

AD HOC Network: A temporary grouping of television stations to carry a specific program or advertising schedule.

Ad Lib: Improvised lines, phrases, or even action bits used by an actor in spontaneous reaction to the given situation of a scene.

AD: Audio Director, person who runs the sound.

AD: See Assistant Director.

ADA: See Audio Distribution Amplifier.

Adaptation: Rewriting of fact or fiction for film presentation, usually in the form of a completed screenplay, or a proposal treatment.

Adapter Rings: A ring that will allow you to use filters that are either too small or too large to normally mount onto that particular camera. Adaptor Rings save money because they allow you to use one filter set on multiple cameras.

Adapter Shell: A holder for the VHS-C cassettes which allows their cassettes to fit into a standard VHS and 8mm players.

Adapter: A connector which allows one type of plug to fit into another type of socket. It's also an audio device that allows a plug of one type to fit a socket of another type.

Adaptive Differential Pulse Code Modulation (ADPCM): A compression technique which encodes the difference between one sample and the next. Variations are lossy and lossless.

ADC: See Analog to Digital Conversion.

Added Scenes: Material, shots, sequences, or scenes written into a script during its principal filming or after its completion.

Additional Camera: An extra camera operator, often needed for complicated action sequence or stunts.

Additional Dialogue Recording: See ADR.

Additional Outlet (A/O): Receivers other than the primary one can be connected to the dish allowing other televisions in the house to be on different programs than the one connected to the primary receiver. An A/O also refers to a convenience outlet where there is not another receiver and the television will show the same programs as the television connected to receiver.

Additional Photography: Focus group or studio reaction

to some shots or scenes may be bad enough to convince the filmmakers to discard them. In some cases, actors are recalled and parts of the movie are refilmed. This is referred to as "Additional Photography," "Reshoots," or "Pickups." Contrast with additional camera, pickups.

Additive Color: In video, the combination of red, green and blue phosphors to generate all other colors.

Address Track: A control/timing track on the edge of videotape which contains control data for quick and accurate location of program material.

Addressable: The technology that enables a cable system to control the networks received by a cable subscriber without scheduling an appointment. Used primarily for pay-per-view programming.

ADI: See Area of Dominant Influence.

Adjacencies: Commercial announcements that are next to (adjacent to, before or after) a program rather than in breaks within the main body of the program.

Adjacent Channel Interference: Wavy lines or two TV images that simultaneously appear on a TV screen. The problem appears when you're viewing a weak television station while another strong television station, one channel number higher or lower, is broadcast nearby or from the same direction as the weak television station.

ADO (Ampex Digital Optics): Trade name for digital effects system manufactured and sold by Ampex.

ADPCM: See Adaptive Differential Pulse Code Modulation.

ADR (Additional Dialogue Recording): Dubbing done in addition to or as a substitution for location sound. The term ADR has a certain appeal, as it obscures the fact that dubbing was involved when it appears in the credits of the film.

ADR: See Automatic Dialogue Replacement.

ADSL: Asymmetric Digital Subscriber Line, a DSL that

sends data quickly downstream (to you) but upstream slowly, allocating the digital resources of the twisted pair efficiently for many download-heavy applications.

ADU: See Audience Deficiency Unit.

Advance: On a composite print, the distance between a point on the soundtrack and the corresponding image. Also an amount given before receipt of services.

Advanced Audio Coder: See AAC.

Advanced Television Systems Committee (ATSC): An organization formed in the early 1990s to set standards for U.S. digital television, including HDTV. The most common formats expected to be used are 480p (525 scan lines, 480 of them active, per frame progressive scan, 720p, and 1080i (1080 scan lines as two 540 scan line interlaced fields).

Advanced Television: Television that will have higher resolution than standard NTSC system.

Advanced Vertical: Special synchronizing signal sent out by a TBC to a VCP to lock the VCP's video playback to the house (TV system's) sync. Makes the VCP play almost in synchronization with the studio cameras.

Advertiser: An individual, business enterprise, or special interest group who uses the media to draw public attention to their products, goods or services.

Advertising Weight: Measure of advertising delivery normally stated by number of commercials delivered, households reached, impressions, gross rating points, etc.

Aerial Image: An actual image that is focused in the air rather than on a card, a screen or a piece of glass.

Aerial Perspective: A shot with visual depth in which distance objects appear less distinctly than objects in foreground, often as a result of atmospheric conditions (e.g., desert heat).

Aerial Shot: Camera shot from the point-of-view of an airborne craft, such as an airplane or helicopter.

AES/EBU: The digital audio standard set by the Audio Engineering Society and European Broadcast Union and used by most forms of digital audio from CDs to D1.

Aesthetic: A philosophy of the beautiful; criteria which define art (or television) as good or bad. Also used to refer to determining factors of television that are neither technological or economic.

AFC Time Constant: An internal circuit design on a TV set which determines how much it jitters and flagwaves when playing tapes.

AFC: See Automatic Frequency Control.

Affidavit: A commercial broadcaster's written document stating a commercial or program aired as ordered.

Affiliate Television Station: A broadcast television station in contractual agreement with a network. The contract grants the network a broadcast signal in the television station's market to air its programs, providing an audience for the network's advertisers. In exchange, the local television station gets "free" programming, and is allowed to air a small number of commercials during these shows. Many television stations also receive a cash payment from the network. See Network Compensation.

AFM: See Audio Frequency Modulation.

A-Frame Edit: A video edit which starts on the first frame of the 5 video frame (4 film frame) sequence created when 24 frame film is transferred to 30 frame video (see 3**:** 2 pull-down). The A-frame is the only frame in the sequence where a film frame is completely reproduced on one complete video frame.

AFT: See Automatic Fine Tuning.

AFTRA: See American Federation of Television and Radio Artists.

AGC: See Automatic Gain Control.

Agency Commission: The commission paid to an

advertising agency by television and radio television stations, networks, and other media for various agency services performed in connection with the placement of advertising in the media. The typical agency commission is 15% of the amount charged by the television station for its time. For example, if the television station invoice is $1000, the advertiser pays the agency $1000 and the agency pays the television station $850. Other arrangements are becoming more common, such as a flat fee or hourly charge for actual services performed.

Agency Meeting: Gathering in which a potential client of an agency meets with the agents to discuss the agents' plan and goals for creating a career for the potential client.

Agent: A manager responsible for the professional business dealings of an actor, director, or other artist. An agent typically negotiates the contracts on behalf of the actor or director, and often has some part in selecting or recommending roles for their client.

Air Check: A recording—film, videotape, audio tape, or transcription—of a program or commercial to check production technique, commercial delivery, etc.

Air Date: Day on which specific program is to be broadcast.

Air Monitor: TV set that shows what is being transmitted or recorded, or, in audio, the speaker that allows an operator to hear being transmitted or recorded.

Air Mouse: Infrared remote control mouse that allows one to operate a computer from some distance away.

Air Quality: Program material that meets technical (generally FCC) standards for quality.

ALC: See Automatic Level Control.

Aleatory Techniques: Techniques in film which depend upon the element of chance. The images are not planned out in advance but are created on the spot by the camera operator. This technique is often used in documentary films.

Algorithm: A formula or set of steps used to simplify, modify, or predict data. Complex algorithms are used to selectively reduce (compress) the high digital audio and video data rates. Algorithms are formulated to selectively reduce the data rate without affecting picture quality.

Aliasing: The undesired jagged or stair-stepped appearance of unfiltered angled lines in an image, graphic, or text.

Alignment Tape: An audio or videotape containing picture and sound reference signals. Used for aligning a tape machine.

All-channel Antenna: Antenna designed to tune in all TV channels.

Allegory: An extended metaphor which often involves characterizing individuals and/or events at representing ideas such as Truth, Beauty, Life and Death.

Alligator Grip: This is a spring-loaded lighting grip used to attach a light to pipes, molding or even doors. The most common place to see these used is in a sound stage or studio. There are special lighting grids made of pipe hanging from the ceiling, allowing you to attach these lighting grips to them.

Allusion: A reference to some well-known event, person or work of art.

Alpha Channel: The fourth channel of a 32-bit RGB image that creates transparency in the image. The other three channels are red, green, and blue.

Alpha Mix: This is a way of combining two images. How the mixing is performed is provided by the alpha channel. The little box that appears over the left-hand shoulder of a news anchor is put there by an alpha mixer. Wherever the little box is to appear, a "1" is put in the alpha channel. Wherever it doesn't appear, a "0" is used. When the alpha mixer sees a "1" coming from the alpha channel, it displays the little box. Whenever it sees a "0," it displays the news anchor.

Alpha Wrap: Tape configuration around the video and audio heads associated with helical scan videotape machines.

Alpha: See Alpha Channel and Alpha Mix.

Alternative Editing Techniques: See Jumpcut, Graphic Match, and Rhythmic Match.

AM: See Amplitude Modulation.

Ambient Color: The shaded color of an object showing darker color where light doesn't hit.

Ambient Lighting: Light that partially fills in shadows, mimicking the light that scatters from everything in the real world. Without ambiance lighting, shadows could get too dark.

Ambient Sound: Background sounds of a particular room or location.

American Cinematic Editors (ACE): Union guild for film/video editors.

American Federation of Television and Radio Artists (AFTRA): The union to which broadcast artists belong.

American Montage: A style of editing, typical of many Hollywood films, which condenses time or summarizes many events in a few shots. Also called Hollywood montage.

American National Standards Institute (ANSI): An organization that, among other things, sets the standards for measuring projector brightness.

American Society of Cinematographers (ASC): An organization founded in 1919 and dedicated to advancing the art of cinematography through artistry and technological progress, to exchange ideas and to cement a closer relationship among cinematographers. Membership is international and by invitation based on an individual's body of narrative film work. Use of the abbreviation ASC, e.g. for on-screen credits, indicates membership in the society. The society publishes "American Cinematographer" magazine.

American Society of Composers and Performers (ASCAP): An agency that licenses the use of copyrighted music.

American Standards Association: See ASA.

A-mode: In this mode, the original footage is edited to the edit master in the sequence of the final program order.

Amp: A measure of the volume of electrical current. Institutional circuits are usually rated for 20A (amps). Electric wires may get hot as this number is approached.

Ampex Digital Optics: See ADO.

Amplifier: Electronic device that makes a weak electrical signal stronger.

Amplitude Distortion: The nonlinear relation of input and output induced harmonics which is a function of voltage fluctuations or power consumption.

Amplitude Modulation (AM): A process used for some radio (AM broadcast) and television video transmission. A low frequency (program) signal modulates (changes) the amplitude of a high frequency RF carrier signal (causing it to deviate from its nominal base amplitude). The original program signal is recovered (demodulated) at the receiver. This system is extensively used in broadcast radio transmission because it is less prone to signal interference and retains most of the original signal quality. In video, FM is used in order to record high quality signals on videotape.

Amplitude: The strength of an electronic signal as measure by its waveform height.

Analog Component Video: Refers to a standard consisting of analog video signals transmitted on three wires, one for luminance (Y), one for the red component from which is subtracted the total luminance (R-Y), and one for the blue component from which is subtracted the total luminance (B-Y). The green component is derived by combining the three components described.

Analog Editor: An editing approach that requires you to

make edits in a set progression like adding links on a chain, as opposed to nonlinear, digital editing which is based on digitized information that can be accessed in a random order.

Analog Monitor: A video monitor which accepts analog signals. Several types of inputs are accepted by analog monitors: composite video, RGB & sync, Y/C, YUV and any combination of these formats. The signals transmitted to an analog monitor are usually between 0 and 1 V and use 75 ohm coaxial cables.

Analog Nonlinear Editor: An editing approach that doesn't digitize the tapes and prepare a final edit from the hard drives, but creates an edit decision list from the timeline on the computer screen. The list later drives the VCPs and VCRs to make the edits.

Analog Recording: A means of recording audio or video whereby the recorded signal is a physical representation of the waveform of the original signal.

Analog Sound: An electronic replica of a sound wave on audio or videotape; the sound wave is converted into an electronic copy or analog. This type of sound recording is being replaced by digital sound. Vinyl albums and audio cassettes create sound through an analog process; compact discs, DVDs and digital audio tape (DAT) store sound digitally.

Analog Video: A video signal made of a continuous electrical signal. The television and VCR can be analog video devices. To be stored and manipulated on a computer, analog video must be converted to digital video.

Analog VTR: Video recorder that records the continuously varying video signal onto the tape (as opposed to digital).

Analog: A method of representing data using continuously varying electrical voltages. Analog video whether transmitted over cables, read from videotapes or broadcast, is subject to degradation due to noise, distortion and other electronic phenomena. Normal signal levels should be within 0.7-1 volt.

Analog-to-Digital Converter (ADC): An electronic device used at the input of digital audio equipment to convert analog electrical signals to digital values whose numbers represent the level and frequency information contained in the original analog signal.

Anamorphic Lenses: These particular types of lenses are used to squeeze the width of an image being shot so that it will fit on a film frame. When the image is projected it is un-squeezed and restored to its original aspect ratio.

Anamorphic: A widescreen film process (under such trademark names as CinemaScope and Panavision) used to create an image wider than conventional television's. The aspect ratio of most films made with anamorphic lenses today is 1:2.35, while the conventional television image's aspect ratio is 1:1.33.

Anchor Point: One of a specified set of logical sector numbers at which descriptor, that identify an extent of a Volume Descriptor Sequence, may be recorded.

Ancillary Data: Data not related to the picture content that is imbedded in the data stream of a digital video signal. Ancillary data is commonly used to transport digital audio, time code, or other picture related auxiliary data.

Ancillary Markets: Any market outside of actual theatrical exhibition. That is, merchandising, production of a book, theatre play, soundtrack, a show about the "Making of" the movie, etc.

Ancillary Rights: These rights enable the screenwriter to receive a percentage of the profits generated by the film in areas such as posters, action figures, books, records, T-shirts, etc.

Ancillary Time Code: BT.1366 defines how VITC and LTC as ancillary data in digital component interfaces.

Angle Menu: DVD Menu used to select the Angle number.

Angle of Acceptance: Vertical and horizontal angle of view of a lens.

Angle of View: This is the size of the field covered by a lens, measured in degrees. However, because of the aperture masks in film, the angle of view for a given lens is generally described in terms of the height and width of a lens.

Angle On: Direction in a shooting script which indicates that another camera angle is to be made of a previous shot, usually to emphasize a specific object in the shot.

Angle: Relationship between the camera and the subject(s) of the shot.

Animatic: Limited animation consisting of art work shot and edited to serve as a videotape storyboard. Commonly used for test commercials.

Animation Stand: Adjustable structure that holds and controls camera and drawings (or objects) for creating animation.

Animation: A form of filmmaking in which inanimate objects or individual drawings are photographed frame by frame with each frame differing slightly from the previous frame. When these frames are strung together and projected at the standard speed of twenty-four frame per second, the result is the objects or drawings to move and hence seem "animated."

Animator: The person who creates the animation.

Anime: A style of animated movie which had its roots in the comic books of Japan. Animation enjoys an immense variety of subject matter and audiences in Japan. Outside of Japan, "Anime" is often used to describe only the adult oriented science fiction and fantasy entries in the field.

Annotation: Comment specifying the source of each script element that is not wholly fictional, including all characters, events, settings, and segments of dialogue.

Announce Booth: A small, soundproofed room or enclosure with a microphone where voice recordings or on-air announcements are made.

ANSI: See American National Standards Institute.

Answer Print: This is the first corrected print made from the A&B Rolls, printed with the optical track. It is sometimes called a married print because it is the first time that picture and sound are wed together on the same piece of print stock.

Antagonist: Character and/or situation that hinders the protagonist from achieving his or her goal(s).

Antenna Booster: Amplifier, attached to antenna wire, used to strengthen a weak antenna signal.

Antenna Joiner: Electrical device which connects to two or more TV antennas and sends the combined signals to the TV set.

Antenna Preamplifier: An electrical device usually connected near the antenna which makes a weak antenna signal stronger.

Antenna Switch: An electrical switch specially designed for antenna signals.

Antenna: Device which either receives or radiates an radio frequency signal.

Anthology: A collection of short segments or stories assembled into a single production.

Anti-alias Filter: A low-pass filter used to bandwidth-limit a signal to less than one-half the sampling rate.

Anti-aliasing: The manipulation of edges (e.g., those between areas with contrasting colors) in an image, graphic, or text to make the edges appear smoother. Anti-aliased edges appear blurred up close but smooth at normal viewing distance. Anti-aliasing is important when working with high-quality graphics for television use.

Anticipatory Shot: Shooting the scene in such a way as to anticipate with the movement of the action by directing the camera where the action is to occur before it occurs.

Anticlimax: Anything that happens in the final few moments of a film that dulls down the story crescendo and leaves the audience feeling let down and unsatisfied.

Anti-flicker Switch: Feature on a scan converter that makes fine lines in computer graphics fuzzier so that they don't flicker when displayed as interlaced video.

Antigravity Hangers: Spring-loaded mechanisms between the lights and the grid to allow the lights to be individually lowered (and stay put at various heights) simply by pulling them down or pushing them up.

Antihalation Backing: Most Kodak, Fuji and Agfa film stocks use an antihalation backing to absorb unwanted light rays. This type of backing is built into the film emulsion to prevent a strong light source from bouncing back through the film, ruining the image.

Antihero: Protagonist who has pronounced personality or character defects or eccentricities which are not usually associated with the hero archetype.

Antikeystone: Feature in some projectors that distorts the projected image so that it looks rectangular on the screen, even though the projector is at an angle to the screen.

Anti-naturalist Performance: Performance style in which the viewer is kept aware that the actor is pretending to be a character.

Aperture (Lens): This is the small hole that opens in the lens and directs the amount of light coming in contact with the film plane. As the Aperture or T-stop number increases the amount of light hitting the film through the gate decreases.

Aperture Correction: Horizontal, vertical or two-dimensional processing of video signal with purpose of correcting frequency response distortions to enhance the sharpness of the picture.

Aperture Delay: The time from an edge of the input clock of the ADC until the time the ADC actually takes the sample. The smaller this number the better.

Aperture Grill: See Shadow Mask.

Aperture Jitter: The uncertainty in the aperture delay. This means the aperture delay time changes a little bit each time, and that little bit of change is the aperture jitter.

Aperture Plates: Mattes on a movie projector, manually adjustable.

Apple Box: This is a wooden box, often helpful on the set to raise up equipment, for the cameraperson to stand on if the tripod is up very high etc. Often you will find them used as seats by the less involved participants. There are also half apples and quarter apples, which as you might expect, are half and one quarter as thick respectively.

Application Format: A specification for storing information in a particular way to enable a particular use.

Approximate Color Consistency: The human ability to visually correct for changes in color temperature. For example, a white piece of paper will appear white under both daylight and incandescent light even though the actual color will vary more than 2,000K.

Arbitron: Formerly American Research Bureau (ARB), an audience market survey (or ratings) service based on diaries distributed to TV viewers and radio listeners (National ARBitron).

Arc: A lighting unit that generates illumination from an electrical arc between two carbon electrodes.

Architecture: In digital video, architecture (sometimes also known as format) refers to the structure of the software responsible for creating, storing, and displaying video content. A architecture may include such things such as compression support, system extensions, and browser plug-ins. Different multimedia architectures offer different features and compression options, and store video data in different file formats. QuickTime, RealVideo, and MPEG are examples of video architectures (though MPEG is also a type of compression).

Archive Footage: See Stock Footage.

Archive Number: A unique number used to label and

match a VCR tape and MO disk cartridge in Quantel archive operations. The same archive number is stored in the tape's user-bits and the MO disk's volume number.

Archive: Long-term storage of information. For editing the requirement is mainly to store the editing decisions and set-ups so that every aspect of the edit can be re-made.

Arcing: A term used in television studio production to refer to the semi-circular sideways movement of the camera.

Area Light: Soft diffused light, like from a fluorescent fixture.

Area of Dominant Influence (ADI): The Arbitron Ratings Co.'s term for the region in which local television stations' signals are dominant. See Designated Market Area.

Arm: A metal rod that is attached to a C-Stand which can extend off to the side.

Arri: Abbreviation for the Arriflex Camera, a lightweight reflex motion-picture camera which comes in both 16mm and 35mm speeds.

Art Department: Crew concerned with visual artistry of a production. Members of art dept. include**:** Art director, assistant art director, draftsman, leadman, production designer, production buyer, property master, set dresser, special effects supervisor, among others.

Art Director: The person who oversees the artists and craftspeople who build the sets. See also production designer, set designer, set director, leadman, and swing gang.

Artifact: In video, refers to something present in the reproduced image, notably crawling dots, rainbow swirls, and color contamination, and that was not present in the original picture or scene. Artifacts are the result of imperfect capture, processing, transmission, storage, and/or decoding of the video signal. With digital video, artifacts can result from overloading the input device with too much signal, or from excessive or improper compression.

Artist Permission: Authorization from a recording artist to use their copyrighted works in a production.

ASA (American Standards Association): This is the sensitivity to light of a particular type of film. It is the specific number used to measure Film Speed. It is the same as IE and ISO. ASA. stands for American Standards Association, the organization that standardized the scale of measurement of film speed.

ASC: See American Society of Cinematographers.

ASCAP: See American Society of Composers and Performers.

Ascender: The part of a letter that rises above the main body, like the top of the lowercase "k."

Aspect Ratio Conversion: Conversion of the T\t picture geometry preserving the scanning format. Note that the video signal itself is aspect ratio independent. A ITU-R BT601 signal may be either 4:3 or 16: 9. Only the display screen has an aspect ratio, thus a 16: 9 signal will appear anamorphic on a 4:3 display.

Aspect Ratio: The proportions of the frame. In 16mm and 35mm the camera photographs a slightly square image, with an aspect ratio of 1.33 to 1. Aspect Ratios are usually shorted to leave out the "- to 1," taking for granted that it will always be in relation to 1, an so "1.33 to 1" can just be called "1.33" In 35mm 1.33 is known as the Academy Aperture. In 35mm the image is usually shot with the Academy Aperture and then masked in the projector to produce a wider image: 1.85 in the U.S. and 1.66 in Europe.

Asperity Noise: Noise that is caused by minute imperfections in the surface of a recording medium (tape). Sounds like a hiss.

Aspherical Lens: Lens formed to a complex shape that provides improved image sharpness, lighter weight, and more accurate color imagery than simple convex and concave lens groups.

Assemble Editing: A method of editing onto blank

(unstriped) tape in a linear fashion. The control track, time code, video, and audio are all recorded simultaneously and joined to the end of the previously recorded material. Major problems can arise. Mixing Insert and Assemble editing in the same session by mistakenly trying to make an insert edit while in Assemble mode will create a hole in the control track at the out point that cannot be repaired. Discontinuities can also be caused if the edit master tape is recorded on more than one VTR. Assemble editing is not relevant to disk-based edit systems; however it still could be of use when playing finished pieces out to tape.

Assemble: The work of organizing and joining the shots of a film in the sequence in which they will appear in the finished print.

Assembly Edit: An edit wherein all existing signals on a tape, if any, are replaced with new signals. See also Insert Edit.

Assignment Editor: The individual in a broadcast newsroom who has the responsibility of assigning news and ENG stories to specific reporters.

Assistant Art Director: An assistant to the art director.

Assistant Cameraman: Member of the camera crew who assists the camera operator and is responsible for the maintenance and care of the camera. Also may do duties of clapper-loader and/or focus puller.

Assistant Director: Sometimes called the "AD" the director's lead hand or foreman. The AD performs many of the routine day-to-day activities on a set including summoning the actors and crew (the Call) and keeping the production on schedule. It is the AD who says at the beginning of a shot "Quiet on the set!" and then orders the camera(s) to "Roll." Duties include tracking the progress of filming versus the production schedule and preparing the call sheets.

Assistant Film Editor: Editing room crewmember responsible for providing any and all required logistical assistance to the editor(s). Duties vary, depending on whether the assistant is working with a picture or sound editor and whether the show is being edited on film or on a nonlinear editing system.

Assistant Production Manager: An assistant to the production coordinator. See also Production Secretary.

Associate Producer: Producer who shares responsibility for creative and business issues.

Association Internationale du Film d'Animation (ASIFA): An organization founded in 1960 in France, chartered under UNESCO, as a membership organization devoted to the encouragement and dissemination of film animation as an art and communication form.

Association of Film Commissioners International: Non-profit educational organization that assists the needs of on-location film, television and commercial production.

Association of Film, Television and Radio Artists: An association with jurisdiction over some works that can be recorded by picture or by sound. See also the Screen Actors Guild.

Association of Motion Picture and Television Producers: Union guild for Film/TV producers.

Associational Editing: The juxtaposition of shots in order to present contrast, comparisons or ideas.

ASV: See Auto Still Video.

ASVOBS Audio Still Video Object Set.

Asymmetric Digital Subscriber Line: See ADSL.

Asymmetrical Balance: Composition arrangement in which a significant object placed close to the center of the picture is balanced by a smaller object some distance from the center.

Asynchronous Sound: Sound which is indigenous to the action but not precisely synchronized with the action.

Asynchronous Transfer Mode (ATM): A method of grouping data into packets and switching them along a route to their destination quickly.

Asynchronous: Any transmission technique that does not require a common clock between two communicating devices, but instead derives timing signals from special bits or characters imbedded the data stream itself.

ATA Carnet: A customs document listing the tools and their origin and destination. It guarantees to a country that they were not bought nor will be sold in that country.

ATAPI (Advanced Technology Attachment (ATA) Packet Interface): An interface between a computer and its internal peripherals such as DVD-ROM drives. ATAPI provides the command set for controlling devices connected via an IDE interface. ATAPI is part of the Enhanced IDE (E-IDE) interface, also known as ATA-2. ATAPI was extended for use in DVD-ROM drives by the SFF 8090 specification.

ATC: See Ancillary Time code.

ATF: See Automatic Track Finding.

ATM: See Asynchronous Transfer Mode.

Atmosphere: Tone or dimension added to the action by concrete or nebulous qualities or elements such as rain, wind, heat, cold, danger, spookiness, tranquility.

A-to-D Converter: An electronic device that converts analog signals to digital. An A-to-D converter is an integral part of digital-video-related technology.

ATR: Audio Tape Recorder.

ATSC A/49: Defines the ghost cancellation reference signal for NTSC.

ATSC A/52: Defines the (Dolby Digital) audio compression for ATSC HDTV.

ATSC A/53, A/54: Defines ATSC HDTV for the USA.

ATSC A/57: Defines the program, episode, and version ID for ATSC HDTV.

ATSC A/63: Defines the method for handling 25 and 50 Hz video for ATSC HDTV.

ATSC A/65: Defines the program and system information protocol for ATSC HDTV.

ATSC: See Advanced Television Systems Committee.

Attenuator: Small electronic device that reduces the strength of an audio signal.

ATV (Advanced Television): The system and the standards defined by the ATSC.

Auction/Bidding War: Situation in which a spec script, book, or written material is presented to several studios, all wanting to buy the work.

Audible Spectrum: When wave motion in the atmosphere has a frequency that lies between 20 and 15,000 Hz it produces a sound which is apparent to the average human.

Audience Deficiency Unit (ADU): Spots added to make up for low or under-delivered ratings.

Audience Duplication: The extent to which an audience of one program or cable network is exposed to a second program or cable network.

Audience Flow: A measure of the change in audience to and between programs. Audience flow shows the percentage of viewers or households who turn on or off a program, switch to or from another channel or remain on the same channel as the previous program.

Audience Fragmentation: A term used to describe the growing number of audience subdivisions. Fragmentation can result from the increase of cable and pay cable television, the growth of independent television stations, VCRs, video games, laser disc players, etc.

Audience Profile: The demographic characteristics of individuals or households reached by a television program or cable network.

Audio Bandwidth: The range of audio frequencies which directly influence the fidelity of a sound. The higher the audio bandwidth, the better the sound fidelity. The highest practical frequency which the human ear can normally hear is 20 kHz. An audio amplifier which processes all frequencies equally (flat response to 20 kHz) and a reasonably high signal to noise ratio, will faithfully reproduce the audio soundtracks of a video recording.

Audio Board: A basic desktop control center used to switch, mix and control audio levels for a variety of audio sources.

Audio Booth: An area—often a separate room—in a production facility where audio signals are controlled and mixed.

Audio Channel: A single audio circuit, generally consisting of an amplifier and one of more audio controls.

Audio Compressor: Audio-processing circuit that reduces dynamic range by simultaneously raising low audio levels and lowering high levels so that a higher average level is achieved.

Audio Console: An electronic mixing and switching device that controls audio sources during a production.

Audio Director: Studio crew member who handles the microphone placement, sound mix, and other audio responsibilities before and during the show.

Audio Distribution Amplifier (ADA): Electronic device that takes in one audio signal and makes several, each as strong as the original.

Audio Dub: VCR feature allowing replacement of the audio signals on a previously recorded tape without disturbing the video signal. When dubbing is not available via the video recorder, audio dubbing can be performed while recording using an audio mixer.

Audio Editing: Similar to video editing. Various portions of audio material are combined and recorded onto the videotape in one continuous form. For example, when a soundtrack is added to a videotape, various sounds such as background music, sound effects and voice narration, may be introduced in order to highlight particular movie scenes.

Audio Expander: An electrical circuit which increases the dynamic range of an audio signal.

Audio Frequency Modulation (AFM): Method of recording hi-fi audio on 8mm and hi8 tapes along with the video. The audio is changed to a varying tone whose vibrations are mixed invisibly with the video vibrations. This technique is used in VHS, SVHS, 8mm and Hi 8 VCRs to record/play hi fi sound, invisibly imbedded in the picture.

Audio Head: Television stationary electromagnet inside a VCR which records the sound on the tape or plays it back. Hi Fi VCRs have audio heads that spin with the video heads.

Audio Insert: An audio dub performed in the midst of an already recorded tape.

Audio Level Control: A volume control. Adjusts sound recording loudness on VCRs.

Audio Levels: Proper audio levels are crucial. If the audio level is too high when recording, overload of the input electronics will cause audio distortion. If audio levels are too low, the signal-to-noise ratio deteriorates. Audio levels are typically indicated either by mechanical VU-meters or electronic LED bar graph meters.

Audio Limiter: Automatic control on a recorder that reduces volume during a recording if the sound becomes too loud. The audio limiter doesn't affect the quiet and medium parts of the recording.

Audio Meter: Meter that indicates the loudness of an audio signal. Could also be a string of LEDs that light up like a bar graph.

Audio Mixing: The blending of two or more audio signals

to generate a combined signal which is often used for audio dub. During video processing, audio mixing may be used to insert narration or background music.

Audio Modulation: Refers to modifying an audio subcarrier with audio information so that it may be mixed with the video information and transmitted.

Audio Monitor: Device that allows you to listen to and check on the quality of a sound signal. Also the switch on a VCR that chooses which channel (or both) is fed to the headphones or in some cases to the VCR's audio output.

Audio Patch Cord: Wire with audio plugs on each end for feeding signals between two audio devices.

Audio Slate: Voice track information recorded at the start of a tape identifying the production. This might include the series title and episode number, the director, the production date and the planned air date. Generally a video slate is also included with essentially the same information.

Audio Speaker: An electro-mechanical device for reproducing audio which changes electrical energy into sound.

Audio Splicing Tape: Adhesive tape used to join the ends of audio recording tape during the editing process.

Audio Still Video (ASV): A still picture on a DVD-Audio disc.

Audio Subcarrier: A specific frequency that is modulated with audio data.

Audio Sweetening: The process of producing optimal sounds by adjusting audio controls to improve the signal-to-noise ratio.

Audio Technician: A person who is in charge of some phase of audio. Generally, a person who operates an audio board.

Audio Track: The portion of the audio videotape that records sound.

Audio Video Interleave (AVI): A data format developed by Microsoft for digital video. Compressed picture and sound data are interleaved in such a way that they proceed synchronous to one another.

Audio/Video Jacks: There are three jacks: one for the video, one for the right channel sound and one for the left channel sound. The audio jacks are necessary for the sound to be in stereo. R/F connectors (coaxial cable) will not provide stereo from the satellite receiver.

Audio/Video Synching: The process of synchronizing the audio data and video images in a combined clip of media.

Audio/Visual Receiver: An audio amplifier with radio tuner(s) (FM stereo and sometimes AM standard broadcast), sometimes a pre-amplifier for a phonograph pickup, tone controls, a number of auxiliary audio and video jacks, and a means of selecting any one of the inputs or audio sources together with the matching video source if any. Generally the selected video signal is not processed in any way; it is simply passed to the video output jack or cable.

Audio: The sound part of a TV broadcast. Sound, turned into an electrical signal.

Audio-1, Audio-2: Names given to the two audio channels on a 3/4U VCR. Audio-2 is often the main channel. Some home VCRs may have two-channel audio or stereo audio.

Audio-follow-video: A special switch that routes an audio signal along with the video signal at the press of a single button, like two switches in one.

Audio-Visual (A/V): Refers to the jacks and connections (if any) on a TV set other than the antenna connections. So named because they were originally used by the audio-visual aids personnel in schools to connect closed circuit equipment, cameras, and videotape recorders.

Audition Channel: An audio monitor circuit enabling a technician to listen to, preview, and cue audio sources before sending them through the audio console.

Audition: The act of checking on a sound signal but not recording it. Also, a mixer channel that can be listened to or adjusted but is not necessarily recorded.

Auteur Theory: A theory popularized by French film critics in the 1950s which argues that the director is the "author" of a film, with artistic control and the power to imbue the work with his or her personal vision.

Authoring: Process of organizing the materials for an interactive disc and putting them into computer language.

Auto Assembly: Process of assembling an edited videotape on a computerized editing system, controlled by an edit decision list.

Auto Fade: Control on some cameras which fades the picture to black at the end of a scene or fades up from black at the beginning.

Auto Iris: System that automatically adjusts a lens aperture to compensate for the brightness of a scene. See Automatic Exposure Control.

Auto Key Tracking: Ability of a video effects system to retain the appropriate size of a keyed-in image when the camera supplying the background video is zoomed in or out. Also called Chroma Key Tracking.

Auto Key: A type of key effect in which one video source is used to establish the external shape of the key.

Auto Preview: Mechanism on a switcher/SEG that automatically displays on the preview monitor any effect not being recorded but ready to be shown once selected.

Auto-assemble: Automatic editing of videotape to conform to previously-generated edit-decision list (EDL).

Autofocus: Electronic system in some cameras that attempts with varying degrees of success to automatically focus subject matter.

Automated Switching: In a television station, the computer-controlled (as opposed to human-controlled) switching between various on-air program and spot sources, according to a previously entered program schedule.

Automatic Dialogue Replacement (ADR): Re-recording of dialogue by actors in a sound studio during post-production, usually performed to playback of edited picture in order to match lip movements on screen.

Automatic Dialogue Replacement Editing: The process of editing sound during Automatic Dialogue Replacement.

Automatic Dialogue Replacement Editor: The person who performs ADR editing.

Automatic Exposure Control: System that responds to varying light conditions by adjusting a camera's iris in an effort to automatically attain optimum overall exposure.

Automatic Fine Tuning (AFT): An electronic circuit in a receiver that "locks onto" a television station and (within limits) doesn't allow it to drift off frequency.

Automatic Focus: Electronic system in some cameras that senses whether the picture is sharp and electrically focuses the lens to correct blurry pictures.

Automatic Frequency Control: A technique to lock onto and track a desired frequency.

Automatic Gain Control (AGC): Circuitry used to ensure that output signals are maintained at constant levels in the face of widely varying input signal levels. AGC is typically used to maintain a constant video luminance level by boosting weak (low light) picture signals electronically. Some equipment include gain controls which are switchable between automatic and manual control.

Automatic Iris Control: Electronic device that automatically alters the lens diaphragm in response to the level of light.

Automatic Level Control (ALC): Circuitry used to automatically adjust the audio recording level to compensate for variations in input volume. Some equipment includes level controls which can switch from automatic to manual control.

Automatic Slating: A camera attachment which records the film shot and take numbers and a visible sync mark which corresponds to a tone fed to an audio recorder.

Automatic Track Finding (ATF): An electromechanical process that eliminates the need for a manual tracking adjustment on a VCR. Also referred to as AST, or automatic scan tracking.

Automatic Transition: A feature on many switchers that automatically executes dissolves, wipes, and other transitions without the use of the fader levers.

Autoplay: A feature of DVD players which automatically begins playback of a disc if so encoded.

Autoscan: Where a multisync monitor or projector can be switched to a selected sweep frequency to match a computer or TV scan rate, an autoscan or multiscan monitor or projector will "sense" the frequency and automatically lock onto it. Also called Multiscan.

Aux: Auxiliary channel in edit decision list (EDL). Also a channel reserved for connecting an external audio and/or video source in a video editing system.

Auxiliary Send: Audio feed that goes to earphone worn by a performer in the studio so that he or she can hear the program signal.

AV: Abbreviation for audiovisual; the making use of or relating to both hearing and sight and to electronic media in general.

Availabilities: See Avails.

Available Light: Light from a natural source or commonly used lamp, as opposed to light added to a scene by using special television lights.

Avails: Commercial breaks that can contain one spot or a cluster of spots. Typically an avail is 30 to 120 seconds in length. It may be subdivided into Unites of 10, 15, 30 or 60 seconds.

Avant Garde: Experimental or highly independent film that is often the forerunner of a new artistic genre.

Average Audience (AA): The number of television households or persons tuning in and viewing a program during an average minute. Typically listed by Nielsen in thousands, the %AA is commonly referred to as a rating.

Average Frequency: The average number of times the reached individual or household has been exposed to the advertising message in a media schedule.

Average Unit Rate: Sometimes referred to as "unit rate," the overall average money earned for each spot sold in a daypart or time period. This figure is calculated by dividing the total dollars by the total number of spots. Example: 10 spots sold for a total of $100 gives an average of $10 per unit.

AVI: See Audio Video Interleave.

AVID Editor: Nonlinear editing system.

Avid: A company that builds digital nonlinear editing suits. Most film producers use them to edit there films before they actually cut their negatives.

A-Wind and **B-Wind:** This is the emulsion position of the film. There are two possibilities, just as there are two sides to a piece of film. Camera original is B-Wind. A print struck from it will be A-Wind. This is because film is printed emulsion against emulsion.

Axis (x, y, z): Used to describe the three dimensional axes available in DVE manipulations. At normal (clear) x lies across the screen left to right, y up the screen bottom to top and z points into the screen. Depending on the power of the equipment and the complexity of the DVE move, several sets of xyz axes may be in use at one time. For example, one may be referred to the screen, another to the picture and a third off-set to some point in space (reference axis).

Axis of Action: In the continuity (or 180) system, the line of action around which the space of the scene is oriented.

Azimuth: The horizontal direction of a satellite. It would have a different azimuth in Los Angeles than in Chicago. Since the satellite is over the equator, wherever you live it will be in the southern sky.

B

B Picture: One of three picture types used in MPEG video. B pictures are bi-directionally predicted, based on both previous and following pictures. B pictures usually use the least number of bits. B pictures do not propagate coding errors since they are not used as a reference by other pictures.

B&W: A term denoting a black and white film or photograph.

B'-Y': The blue-minus-luma signal, also called a color difference signal. When added to the luma (Y') signal, it produces the blue video signal.

Baby Boom: Small boom stand for holding a microphone.

Baby Legs: A short tripod.

Baby Plate: A steel plate with a baby pin (5/8 inch pin) welded on to it. These plates are used for mounting lights or grip heads on a wall, box, or other surface.

Baby: Usually a reference to a 1K light unit. It is also used to describe any light unit which is smaller than a standard size unit of comparable intensity (i.e. baby 1K, baby 2K, baby 5K, etc.). For grips, it refers to anything with a 5/8 inch stud (i.e. baby plate).

Back Channel: A means of communication from users to content providers. Today, a common type of back channel is an Internet connection using a modem.

Back Focus: The distance between the lens and the pickup chip. To remain in focus while zooming, the lens' back focus must be adjusted precisely. Also, the act of adjusting a lens' back focus.

Back Light: In the three-point lighting system, the source of illumination placed behind and above the actor. Its main function is to cast light on the actor's head and shoulders, creating an outline of light around the actor to distinguish him or her from the background.

Back Lot: This term refers to the exterior sets of typical locals such as a frontier town, a city street, a European Village which were built on the back lots of the big Studios during the studio era.

Back Porch: The portion of the video waveform between the trailing edge of the horizontal sync and the start of active video.

Back Projection: A photographic technique whereby live action is filmed in front of a screen which the background action is projected on. Originally used for scenes occurring in vehicles. Contrast this with a matte shot. Also called Rear Projection.

Backdrop: Artificial background, usually painted on a cyclorama, curtain, or flats, used to achieve the effect of a natural environment such as a forest, beach, prairies, mountains, or other landscape in a shot or sequence.

Backend: Film's profit from theater ticket sales, video rentals, and ancillary markets.

Background Artist: Individual responsible for designing or constructing the art placed at the rear of a set.

Background Generator: SEG circuit that adds color to a black background, useful for keying words onto a colored background.

Background Light: Any light used to illuminate the background of a set or scene, but not intended to illuminate the subject.

Background Music: Soundtrack music intended to enhance drama or visual elements during a production.

Background Projection: See Rear Projection.

Background: This is a term with a broad range of meanings, depending upon the context. In production, it has the same connotation as "atmosphere," meaning extras that are staged to supply detail in the form of normal human traffic in a scene. In sound, it can mean the same as 'ambience' or it may refer to relative volume. When used in the discussion of composite photography, the Background is the component of the image that is the opposite of foreground. This would include any non-primary elements such as scenery, walls, crowds of people, etc.

Backhaul: The act of sending a program or newsfeed via satellite from a local area back to the main distribution area for rebroadcasting via satellite.

Backing Track: Prerecorded accompaniment for a singer or voiceover actor who then listens through headphones to a replay as he/she performs. Generally, the two signals are ultimately mixed to produce the final recording.

Backlight Compensator (BLC): A camera control that opens up the lens aperture two-to-three F-stops. Used to compensate for the error made by automatic iris circuits when shooting into light or against a bright background.

Backlight Switch: Camera control which overrides auto iris system and opens the iris two or three F-stops. Commonly used in backlit situations where auto-iris would result in dark skin tones and underexposure.

Backlot: A large, undeveloped area on studio property used for constructing large open-air sets or for filming wilderness scenes.

Backspace: Act of moving a videotape backward slightly. Helpful in producing glitch-less (clean, smooth) edits. It also means to move a tape backwards a ways and park it in preparation for an edit or to give the tape space for the preroll.

Backtiming: A method of calculating the in-point by subtracting the edit duration from the known out-point to determine when segments should start in order to bring the show out on time.

Back-to-Back: A program or commercial immediately preceding or following another.

Back-up Schedule: A "plan B" schedule of shots or scenes that can be substituted for planned shooting schedule in case of illness of a principal actor, bad weather, mechanical or technical problems, etc.

Backwind: Rewinding film in the camera to shoot a Double Exposure.

Balance Stripe: A second stripe found on 35mm stripe mag stock and super-8 sound film to prevent warping.

Balance: In video and film, the blending of three colors (red, green, and blue in video; yellow, magenta, and cyan in film) to produce a spectrum of colors. Different video processes and film stocks favor some colors over others, resulting in various types of color balance.

Balanced Audio Input: An audio input into a device that consists of two conductors surrounded by a grounded shield.

Balanced Cable: In audio systems, typically refers to a specific cable configuration that cancels induced noise.

Balanced Line: An audio cable with three wires, two inside a shield.

Balanced: An audio circuit with 3 wires, two which carry signal, and a third which is connected to a ground (grounded).

Ballast: An electrical transformer that properly conditions the electrical power to run HMI lights.

BALUN (Balanced Circuit to Unbalanced Circuit Transformer): A device used to connect a circuit (unbalanced) one of whose two conductors is grounded to a two conductor circuit (balanced) neither of whose conductors is grounded. A typical example of use in video products is a 300 ohm antenna cable (balanced) connected to a 75 ohm antenna cable (unbalanced).

Band Separator/Joiner: Electrical device which separates combined bands (like VHF, UHF, FM) into separate bands (like FM alone) or combines separate bands so the signal can travel on a single cable.

Band: A range of radio frequencies used for a certain type of communications. Also a set of related frequencies. UHF (ultrahigh frequency) is one band 470-890 MHz (megahertz).

Banding: A picture artifact or fault whereby smooth brightness or color gradients appear to be comprised of bands of brightness or color, often the result of too few bits used to represent each sample of a picture. Banding could make a billiard ball look like a sliced onion.

Bandpass Filter: The portion of the video waveform between the trailing edge of the horizontal sync and the start of active video.

Bandpass: The range of wavelengths over which a component will meet specifications.

Bandwith: The amount of data, measured usually in bits per second, that can be sent through a dedicated (leased) transmission circuit.

Bankable Star: A well-renowned actor/actress whose presence in a film is enough to guarantee finance and improve its gross takings.

Banned: Many countries have either government or official movie classification boards who are responsible for determining the suitability of a movie for release in their country or region. These boards occasionally block the release of a movie either in theaters or on video. Often, a banned movie will find its way around a ban by means of bootleg distribution. See also Censorship.

Bar Sheets: A chart showing words of dialogue which have been recorded, and the number of motion picture frames of duration for each syllable.

Barndoors: Metal flaps on a lighting instrument that can be closed or opened to direct the light, and shade areas where light is undesirable.

Barney: A quilted cozy that fits around a camera to reduce camera noise. Generally it is only effective on a camera that is pretty quiet to begin with. The term comes from barney blanket, a kind of horse blanket.

Barrel Connector: An adapter with a socket at each end which allows two cables to be connected together.

Bars and Tone: A carefully-controlled audio and video signal generally recorded at the beginning of a videotape and later used to properly setup playback equipment. "Bars" consist of a test pattern of color bars "tone" consists of an audio signal of a certain frequency at 0dB.

Barter Syndication: A program that can be aired by the TV or radio television station free of charge or for a reduced charge. The program supplier earns revenue by reserving some of the Avails in the program for its own commercial use. Sometimes additional spots sold by the program supplier are aired outside the bartered program (see Spins). In "cash plus barter," the syndicator also receives some money from the television station in addition to the airing of the barter Spots.

Barter: A method of buying media (generally local TV) whereby the cost of the media is paid by the advertiser by merchandising or services as opposed to cash.

Base Light: Even lighting used over a set or production area. Often a set is initially lit with basic (base light) illumination, and then the key and back lights are added.

Base Television Station: Central digital control system used for adjusting levels, color balance, etc. of several studio cameras.

Base: In film production terms, the celluloid backing of a piece of film to which the emulsion adheres. Film has two basic elements: The base is the clear, perforated strip, and the emulsion is the thin, light-sensitive layer that is glued onto it.

Baseband: Any signal in its original form prior to modulation or after demodulation.

Base-to-Base Splice: A splice made with the base side of the end of one piece of film overlapping the base side of the piece to which it is being attached.

Basic Cable: Channels received by cable subscribers for no additional charge. Basic channels are usually advertiser supported (commercial carrying) and/or the cable company pays a per-subscriber fee to the programming service.

Basic Level Videodisc Player: Like a movie, this videodisc and player can only start at the beginning and play to the end of a program. There is no interactivity.

Basic Make-up: Primary or elemental make-up. As opposed to corrective or character make-up, make-up intended to significantly alter or enhance appearance.

Bass: Audio tones of low frequency, typically below 300Hz.

Battery Light: A battery operated, portable light commonly used for ENG work. Often mounted on top of the camera.

Bayonet Mount: A camera mount that locks a lens securely into place with only a partial turn. This contrasts with the C-mount that requires numerous turns to secure. This connection is popular on professional cameras.

Bazooka: Similar to a 2K stand, but without support legs. It has a junior hole at one end and a junior stud at the other, and it usually has a sliding riser.

BBC: British Broadcasting Corporation.

BCA (Burst Cutting Area): A circular section near the center of a DVD disc where ID codes and manufacturing information can be inscribed in bar-code format.

BCU: See Big Close-up.

Beaded Screen: Projection screen covered with tiny glass beads (looks like white sandpaper). It has a gain of 2 or 3.

Beam Projector: A spotlight that projects a focused, hard and generally narrow beam of light often used to simulate sunlight coming through a window.

Beam Splitter: A prism or diachronic mirror device used behind a color TV camera lens to divide the light into the three primary colors.

Beam: A satellite transmission pattern. It may be wide, narrow or spot. This affects the satellites footprint.

Beat: An assigned area covered by reporter on a regular basis. Beat is also a term used to describe a periodic variation of amplitude resulting from the addition of two slightly different frequencies.

Beauty Shot: Identifying close-up shot of an advertised product.

Beaver Board: A 2K pigeon on an apple box.

Bed: Background music used underneath a narrator or foreground dialog. Primarily applied to commercial radio or television spots.

Beef: The output of a light.

Beefy Baby: A heavy duty 2K stand without wheels.

Beep: A short duration soundtrack tone aligned to a point on the film for precise reference in synchronization in the editing and printing processes.

Behind the Scenes: The off-camera goings on associated with film/videomaking.

Below-the-line Costs: Ongoing costs realized whether a production company is doing a show or not. Overhead. Examples: staff engineering and production personnel, equipment amortization, telephone, taxes. Also called Below-the-line Budget and Below-the-line Expression.

BER (Bit Error Rate): Bit errors are caused by interference, or loss of signal, so the stream of bits composing the DTV picture is disrupted. A measure of the errors in a transmitted signal.

Best Boy: The chief assistant, usually of the gaffer, but more often lately used as a general term for the second in command of a group. This term is likely borrowed from early sailing and whaling crews, as sailors were often employed to set up and work rigging in theatres. There are no "best girls" per se; female chief assistants are also called "Best Boys."

Best Light: Similar to a One Light, but by implication, the timer has gone through the film more thoroughly in selecting a timing light that will agree with the majority of the footage.

Betacam SP: A superior performance version of Betacam. SP uses metal particle tape and a wider bandwidth recording system.

Betacam: A tape format and transportable combination camera and recording (camcorder) system developed by Sony and introduced in 1982. Betacam uses a variation of the Y, R-Y, B-Y analog component format.

Betamax: A consumer-quality 1/2-inch videocassette format developed by Sony which eventually was supplanted by VHS.

B-Film: A low budget film shot with second rate actors.

B-Frame: In a video compression scheme, a frame for which enough data is stored so that overlaying the data atop either the complete frame after or the complete frame before will yield that frame. Useful for single step backwards. See also I-Frame, P-Frame.

Bias Light: A dim light inside a camera directed toward the front surface of the CCDs or pickup tubes intended to increase camera sensitivity while controlling video noise.

Bicycling: Shipping a television program or commercial (tape or film) from television station to television station for on-the-air use, rather than sending each television station its own individual copy. Each television station is responsible for sending the program to the next television station on the list in a timely manner. Today satellites have replaced bicycling as the primary source of transmitting programs and commercials.

Bidirectional Microphone: A microphone designed to pick up sound on two sides of the microphone along the same axis, but reject sound from any other direction.

Bidirectional Prediction: A form of compression in which the codec uses information not only from frames that have already been decompressed, but also from frames yet to come. The codec looks in two directions: ahead as well as back. This helps avoid large spikes in data rate caused by scene changes or fast movement, improving image quality.

Big Close-up (BCU): A shot taken very close to the subject (closer than would be necessary for a close-up), revealing extreme detail. (i.e., part of the human face).

Billboard: A brief announcement, usually 3, 5 or 10 seconds in length, and usually earned by advertisers paying extra for the program being ordered. Billboards are afforded, in most instances, at the top and bottom (beginning and end) of the show. The product and/or sponsor's name is mentioned in a statement such as "…the following portion of (program) is being brought to you by (sponsor)…."

Billing: A great deal of importance is placed on the relative sizes, positions, and order of names and the movie's title in printed publicity material as well as the opening credits. Generally, higher positions designate higher importance. Additionally, there is significant given to names which appear before or above the actual title of the movie. The person whose name is shown first in the credits or whose name is at the top of an advertisement is said to have received "top billing." If more than one name appears at the same time or at the same height, they are said to have "equal billing," with the importance of the people concerned decreasing from left to right.

Bin: A reference to a storage container lined with a cloth bag, into which cut film or sound stock may be arranged and hung. In digital audio and video terms, this can be related to a film and/or directory from which stored shots or sound segments are selected for use. See also Trim Bin.

Binary: A numbering system using the digits "0" and "1" in the decimal system. All computer programs are executed in binary form.

Biographic Picture: A filmed story of a person's life story.

BIOS: (Basic Input/Output System): Computer device that manages communications between the basic computer and peripherals, such as the keyboard and monitor.

BIOS: See Basic Input/Output System.

Bi-Phase: Electrical pulses from the tachometer of a telecine, used to update the film footage encoder for each new frame of film being transferred.

Bird: A satellite. Sometimes used as a verb to indicate the transmission of a video program or segment.

Bird's Eye View: A view of a scene photographed from directly overhead; an overhead shot.

Birefringence: An optical phenomenon where light is transmitted at slightly different speeds depending on the angle of incidence. Also light scattering due to different refractions created by impurities, defects, or stresses within the media substrate.

Bit Error Rate: See BER.

Bit Part/Role: Small unimportant role, usually lasting only one scene.

Bit Rate: The amount of data transported in a given amount of time, usually defined in Mega (Million) bits per second (Mbps). Bit rate is one means used to define the amount of compression used on a video signal. Uncompressed D1 has a bit rate of 270 Mbps, DV is 25 Mbps, etc.

Bit Speed: Speed at which units of digital information can be transmitted from one point to another.

Bit Stream Recorder: A device capable of recording a stream of digital data but not necessarily able to process the data.

Bit Stream: A continuous series of bits.

Bit: A minor acting role in which an actor may only speak a few lines. Also a single element (1 or 0) of digital information.

BITC (Burned-In Time Code): The time code information is displayed within a portion of the picture, and may be viewed on any monitor or TV.

Bite: Sound bite. A short interview segment with associated audio intended as part of a production.

Bitmap: An image made of a two-dimensional grid of pixels. Each frame of digital video can be considered a bitmap, although some color information is usually shared by more than one pixel.

Bits Per Pixel: The number of bits used to represent the color or intensity of each pixel in a bitmap. One bit allows only two values (black and white), two bits allows four values, and so on. Also called color depth or bit depth.

Black a Tape: The process of recording a blackburst signal across the entire length of a tape. Often done before recording edited footage on the tape to give the tape clean, continuous video and sync and to insure there is no video already on the tape.

Black Bag: A black changing bag functions as a portable darkroom. You can use a black bag to load a film magazine. This bag is basically two bags, one inside the other and is used from everything from loading film to fixing camera jams.

Black Balancing: Electronically setting the black level of a camera to correspond to TV black.

Black Box: A term used to describe a piece of equipment dedicated to one specific function, also called dedicated hardware.

Black Clipping: A circuit used to limit the black level of video so that it does not intrude into the sync.

Black Comedy: A comedy in which the humor is derived from subjects which are typically considered "serious," or for which humor is usually considered as unsuitable. Common examples are death, war, suffering, and murder.

Black Leader: Black, opaque film, often specifically called black emulsion leader. It is what the negative cutter uses when preparing A&B rolls. It is very important that it be emulsion leader rather than plastic leader when used for A&B rolls, since plastic leader cannot be cement spliced. It also must be very opaque, not any black piece of film will do.

Black Level: The level of the video signal corresponding to the maximum limits of the black areas of a picture.

Black Set: A reference level for black balance adjustment.

Black Wrap: Black Aluminum foil which is used for wrapping lights, to control light spill, and for making small flags.

Blackburst: A composite color video signal comprised of sync, colorburst and black video. It is used to synchronize (genlock) other video sources to the same sync and color information. Blackburst generators are used in video studios to "lock" the entire facility to a common signal ("house sync" or "house black").

Blacklist: A list of filmmakers or actors who have either been formally or informally discriminated against, due to their personal, political, social, or religious beliefs. In 1950s America, McCarthyism resulted in numerous filmmakers being blacklisted.

Blackout: The fading of a video signal to black to indicate, for example, the end of a show. In broadcasting, it means a particular programming service may not be available in certain areas of the country. Usually because of contractual agreements.

Black-Video-Black: See BVB.

Blanked Tape: A videotape intended for insert editing onto which black (or color bars) and a control track have been recorded. Also called a Crystal Tape.

Blanket Contract: A type of contract used when an agency or client makes a contractual agreement to run a certain amount of spots during a fixed period of time, for a given amount of money, even though the exact schedule or product is not known at the time of signing. Also called Bulk Orders or Master Contracts with Insertions.

Blanking Interval (Horizontal & Vertical): The horizontal blanking interval is the time between the end of one scanning line and the beginning of the next. The vertical blanking interval is the time between the end of one video field and the beginning of the next. Blanking occurs when a monitor's electron beam is positioned to start a new line or a new field. The blanking interval is used to instantaneously reduce the beam's amplitude so that the return trace is invisible. See Vertical Interval Switching.

Blanking Level: Also known as the pedestal, it is the voltage level produced at the end of each horizontal picture line which separates the portion of the video signal containing the picture information from the portion containing the synchronizing information. This voltage makes the electron beam "invisible" as it moves to draw the next visible line.

Blanking Pulse: Video signal which blacks out a video image between successive scanning lines, fields, and frames. See Blanking Interval.

Blanking: The part of the video signal that contains no picture information. Used for synchronizing, time code, closed captions, etc. On a CRT display, the scan line moves from the left edge to the right edge, jumps back to the left edge, and starts out all over again, on down the screen. When the scan line hits the right side and is about to be brought back to the left side, the video signal is blanked so that you can't "see" the return path of the scan beam from the right to the left-hand edge. To blank the video signal, the video level is brought down to the blanking level, which below the black level if a pedestal is used.

BLC: See Backlight Compensator.

Blender: Video mixing/keying/color correcting facility in Quantel's Henry.

BLER (Block Error Rate): A measure of the average number of raw channel errors when reading or writing a disc.

Blimp: A soundproof camera housing which prevents the noise of the camera's motor from being recorded on the soundtrack.

Blimped Camera: The term is used not to mean a camera in a blimp, but a camera that is designed with internal soundproofing without the need for an external blimp. For instance, with an Arri BL the "BL" stands for "blimped."

Blip Tone: A sync pop. This is usually done by placing a piece of 'toned' tape on a particular frame of film to establish a 'synch' point.

Block Error Rate: See BLER.

Block Programming: A series of programming with common demographic appeal scheduled one after another.

Blockbuster: A movie which is a huge financial success.

Blocking: The actor's movement around a set; the director's incorporation of the actor into the mise-en-scene. In any scene where there is movement it is first necessary to block out or choreograph all movement and then rehearse it so that the camera and microphone can stay with the performer and also so that no unwanted images get caught on film. Also a term referring to the occasional blocky appearance of compressed video (an artifact), caused when the compression ratio is high enough that the averaging of pixels in 8x8 blocks becomes visible.

Blonde: An open face 2K lighting unit, also known as a "mighty."

Blooming: A dark halo that surrounds bright areas in a scene caused by aberrations in the video process. Generally associated with tube-type cameras.

Bloop: The sound produced in an energized amplifier and speaker system when a film splice passes the photo cell to which the amplifier is connected.

Blooper: An error in speech made by on-camera talent.

Blooping Tape: Tape used to cover unwanted portions of soundtracks.

Blooping: Any method of silencing unwanted noise produced by the passage of a splice through a sound reproducer. Traditionally, a small opaque tape is fixed over a splice.

Blow Down: The actual term for the opposite of a blow up is a Reduction Print, but this term has been coined by Colorlab in Rockville, Maryland, for a reduction print made from super 16mm to regular 16mm, as an alternative to the much more expensive process of blowing up super 16mm to 35mm.

Blow Up: An optical enlargement of a film from one gauge to another, such as 16mm up to 35mm. The opposite of a blow up is a Reduction Print.

Blue Book: The document that specifies the CD Extra interactive music CD format (see also Enhanced CD). The original CDV specification was also in a blue book.

Blue Pages: Revised pages (of any color) that are substituted for existing script pages.

Blue Pencil: To edit or censor a script.

Blue Screen Process: Process of photographing action in front of a blue screen to make a matte special effect possible. On a newsroom set, areas of the background that are blue (or green), onto which live images or maps are substituted through the Chroma Key process.

Blue Sky: The potential for the movie to make a greater profit than that which is guaranteed by distribution and pre-sale agreements.

BMI: Broadcast Music, Inc. Agency that licenses the use of copyrighted music.

B-Mode Edit: An editing method where the footage is assembled in the order it appears on the source reels. Missing scenes are left as black holes to be filled in by a later reel. Requires fewer reel changes and generally results in a faster edit session.

BMP: See Bitmap Image.

BNC Connector: A type of connector used on some VCRs, video and RF equipment providing twist-lock capability. The jack has a cylindrical "bayonet" shell a little larger around than a pencil (about 3/8 inch) and a center hole that a paper clip wire would just fit (1/32 inch). The plug has a thin matching center pin which may be the center conductor itself of a coaxial cable. The outer shell of the plug fits over the jack shell and is locked in place by a quarter turn twist to engage protrusions on the jack shell.

Boarder Merger: A composition problem in which subject matter is awkwardly and inappropriately cut off at one edge of the video frame.

Bob: Method of line doubling where an intervening scan line is synthesized (its content interpolated) using its neighbors. So named because under some conditions the image appears to vibrate (bob) up and down. See also Weave.

Bobbinet: Black mesh cloth which is used for grip scrims. It also is available in rolls for darkening windows.

Body Brace: A type of camera mount which attains some measure of stability by resting on the shoulder and attaching to the chest or belt area.

Body Wash: Body make-up used by actors.

Bolex: One of the more widely used 16mm non-sync cameras, it is made in Switzerland by the Paillard Company. There are many varieties, non-reflex, reflex, spring-wound and electric motor driven. But when someone says "Bolex," typically they mean a reflex, spring-wound model, such as the Rex-4.

Bonus Spot: A spot given to a buyer free of charge to make up for low ratings or as a consideration for buying other spots. Some bonus spots carry value in that they MUST air as part of the contract, while others are not committed as part of the deal: if they don't run, they are not made good.

Book A: The document specifying the DVD physical format (DVD-ROM). Finalized in August 1996.

Book B: The document specifying the DVD-Video format. Mostly finalized in August 1996.

Book C: The document specifying the DVD-Audio format.

Book D: The document specifying the DVD record-once format (DVD-R). Finalized in August 1997.

Book E: The document specifying the rewritable DVD format (DVD-RAM). Finalized in August 1997.

Bookend: Two usually related commercials, one airs at the beginning of a commercial break and one ends at the end of the commercial break.

Boom Operator: A member of the sound crew who operates the boom microphone. See also Sound Recordist.

Boom Shot: A continuous single shot incorporating any number of camera levels and angles; Hitchcock used this technique for almost all of the filming in The Rope. See also Dolly Shot.

Boom Up/Boom Down: To raise or lower the height of a camera mounted on a boom or pedestal.

Boom: A pole used to extend a microphone above a film set, permitting synchronous sound recording without interfering with actors. Also called a Boom Microphone.

Booster Light: Lamps, usually arc lights, used during exterior daylight shooting to illuminate shadowy areas.

Bootleg: An unofficial and illegally copied or distributed version of a movie, often of a substandard quality. Bootleg videos are often available for movies that have yet to be released in a particular country, or have been banned.

Border: The boundary between two merged video pictures, as created with chroma key or wipe effects.

Bottom Chop: A flag or cutter which is used to keep light off of the floor or the lower part of a scene.

Bottom of the Hour: Refers to the half-hour (as on a clock face); e.g. 12**:** 30, 3**:** 30, etc.

Bounce Card: A white or silver card used for soft indirect lighting of the subject by bouncing light off the card. Can also be used to provide a gentle brightening of shadow areas. Especially out-of-doors as it does not require power.

Bounce Light: An even, diffused light over an area that results from light being reflected off of the ceiling or wall in a room.

Bounding Box: A cube drawn around an object which defines the extents of the model. Often used in collision detection if the bounding boxes of two objects have touched, the objects have collided.

Box Office Gross: The total amount of money paid by the public for cinema tickets to the movie. Breakdown Sheet Part of the Shooting Schedule, the Breakdown Sheet contains all the important information about each scene (except for those determined by the director). Displays length, description, cast, makeup and other special requirements to keep everyone informed. Call Sheet A sheet which notes which scenes will be shot the next day, and the basic cast, crew and special requirements for each scene, as well as scene times and call times.

Box Rental: A fee or allowance paid to a crewmember for providing his/her own equipment or other specialized apparatus for use in a production.

BPS (Bits per Second): A unit of data rate.

Bracketing: The filming of several takes of the same shot at different F-stops to achieve the desired result. Usually this technique is applied to shooting titles much more than anything else.

Branch Holder: A pipe-like unit with a locking nut which is used to hold branches, wooden poles, or other items.

Breakaway: An edit where the audio and video do not automatically switch together, but one comes before or after the other. Often audio will come in before the corresponding scene starts. Also called Split Edit or L-cut. Breakaway is also a prop used during action sequences that looks real but is constructed so that it will break easily without harm to performers.

Breakdown: Analysis of script segments for the purpose of determining best sequence for shooting scenes and setups. Also referred to as Breakdown Sheets. May also refer to analyzing a script for a cost analysis.

Breakup: A brief interruption of video caused by a technical problem.

Breast Line: A guide line attached to anything being hauled up on a crane or by a pulley.

Brechtian Performance: Anti-naturalist, confrontational performance style based on the theories of German playwright Bertolt Brecht. He demanded that the viewer constantly be made aware of the fact that he or she is watching a play and that he or she should be distanced from the characters.

Breezeway: The portion of the video waveform between the trailing edge of horizontal sync and the start of colorburst.

Bridge: Usually a short transition spot (perhaps as long as 10 seconds) between two commercials. Time and television station ID, a short PSA, or a brief weather update could be a bridge.

Bridging Shot: A shot (cut) used to cover a break in time or other break in continuity.

Brightness Control: A control used to adjust the illumination of viewfinders, monitors, and receivers, but not affecting signal levels from cameras or other picture sources.

Brightness: This refers to how much light is emitted by the display, and is controlled by the intensity of the video level.

Broad: A broad-beam floodlight with a rectangular reflector used to light backgrounds and to throw light over a wide set area.

Broadband: The transfer of data packets at rates higher than traditional dial-up connections (i.e., ISDN, 56K modem). Broadband streams enable the transmission of a larger and higher-quality image and richer and better-quality sound. Broadband streams are available with DSL, cable modem, T-1 or higher Internet connections.

Broad-beamed Lamp: A lighting instrument within a rectangular reflector that creates a broad, somewhat diffused light.

Broadcast Calendar: A universally agreed upon calendar, used in accounting procedures in the broadcast industry, containing four or five weeks, with each month beginning with the first Monday of the calendar month.

Broadcast Quality: A nebulous term used to describe the output of a manufacturer's product . Usually at least means that the technical specifications meet the FCC rules for broadcasting.

Broadcast Standards and Practices (BSP): The units within TV networks that make sure offensive material is not broadcast: TV's internal censors.

Brute: A brute arc light, usually 225 amps DC powered.

BS.707: This ITU recommendation specifies the stereo audio specifications (Zweiton and NICAM 728) for the PAL and SECAM video standards.

BSP: See Broadcast Standards and Practices.

BST-OFDM (Bandwidth Segmented Orthogonal Frequency Division Multiplexing): Process that attempts to improve on COFDM by modulating some OFDM carriers differently from others within the same multiplex. A given transmission channel may therefore be "segmented," with different segments being modulated differently.

BT.1119: Defines the Widescreen Signaling (WSS) information for NTSC and PAL video signals. For (B, D, G, H, I) PAL systems, WSS may be present on line 23, and on lines 22 and 285 for (M) NTSC.

BT.1124: Defines the Ghost Cancellation reference (GCR) signal for NTSC and PAL.

BT.1197: Defines the PALplus standard, allowing the transmission of 16**:** 9 programs over normal PAL transmission systems.

BT.1302: Defines the transmission of 16**:** 9 BT.601 4**:** 2**:** 2 YCbCr digital video between pro-video equipment. It defines a parallel interface (8-bit or 10-bit, 36 MHz) and a serial interface (360 Mbps).

BT.1303: Defines the transmission of 16**:** 9 BT.601 4**:** 4**:** 4**:** 4 YCbCrK and RGBK digital video between pro-video equipment. Two parallel interfaces (8-bit or 10-bit, 36 MHz) or two serial interfaces (360 Mbps) are used.

BT.1304: Specifies the checksum for error detection and status for pro-video digital interfaces.

BT.1305: Specifies the digital audio format for ancillary data for pro-video digital interfaces.

BT.1358: 720 x 480 (59.94 Hz) and 720 x 576 (50 Hz) 4**:** 2**:** 2 YCbCr pro-video progressive standards.

BT.1362: Pro-video serial interface for the transmission of BT.1358 digital video between equipment. Two 270 Mbps serial interfaces are used.

BT.1364: Specifies the ancillary data packet format for pro-video digital interfaces.

BT.1365: Specifies the 24-bit digital audio format for pro-video HDTV serial interfaces.

BT.1366: Specifies the transmission of time code as ancillary data for pro-video digital interfaces.

BT.470: Specifies the various NTSC, PAL, and SECAM video standards used around the world. SMPTE 170M also specifies the (M) NTSC video standard used in the United States. BT.470 has replaced BT.624.

BT.601: 720 x 480 (59.94 Hz), 960 x 480 (59.94 Hz), 720 x 576 (50 Hz), and 960 x 576 (50 Hz) 4**:** 2**:** 2 YCbCr pro-video interlaced standards.

BT.653: Defines the various teletext standards used around the world. Systems A, B, C, and D for both 525-line and 625-line TV systems are defined.

BT.656: Defines a parallel interface (8-bit or 10-bit, 27 MHz) and a serial interface (270 Mbps) for the transmission of 4:3 BT.601 4**:** 2**:** 2 YCbCr digital video between pro-video equipment.

BT.709 This ITU recommendation specifies the 1920 x 1080 RGB and 4**:** 2**:** 2 YCbCr interlaced and progressive 16**:** 9 digital video standards. Frame refresh rates of 60, 59.94, 50, 30, 29.97, 25, 24, and 23.976 Hz are supported.

BT.799: Defines the transmission of 4:3 BT.601 4**:** 4**:** 4**:** 4 YCbCrK and RGBK digital video between pro-video equipment. Two parallel interfaces (8-bit or 10-bit, 27 MHz) or two serial interfaces (270 Mbps) are used.

BTA (Best Time Available): In broadcasting, a phrase that usually refers to a sales plan offering the advertiser the best commercial positions that are left after the specified times and other higher priority spots have been placed. See also ROS.

BTSC: A technique of implementing stereo audio for NTSC video. One FM subcarrier transmits a L+R signal, and an AM subcarrier transmits a L-R signal.

Buddy Film: A film genre most popular in the 1970's which focused on male characters and their stories.

Budget Tracking: A process whereby the actual expenses in a production are regularly compared to originally-projected expenses.

Budget: Funds required to produce a film or television production, derived by combining all projected expenses for equipment, salaries, locations, travel, and all other above-the-line and below-the-line production costs.

Buffer: An area of internal storage (temporary holding area) that compensates for a difference in the rate of flow of data, or time of occurrences of events, when transmitting data from one part of a computer system to another. Also an area of storage, temporarily reserved for use in performing an input or output operation, and into which data is read or from which data is written.

Bug: A mistake or malfunction in the computer software that causes inaccurate or inappropriate results when accurate data is processed. The term derives from the early days of computing when systems were created with mostly vacuum tubes. When researching an unexpected result, the engineers discovered a moth between the contacts of two separate tubes, causing a short circuit.

Build Up: Creating tension in a dramatic production by arranging scenes and selecting music that build to a crisis or climax.

Bulk Erase: To completely erase an audio or videotape by the use of a strong electromagnet.

Bump Down: Copying video information from higher-quality VTR format to a lower-quality one.

Bump Up: To copy segments from one videotape to a larger or higher quality videotape format.

Bumped Spots: Spots that are ordered by an advertiser, but did not or will not air for various reasons. Examples**:** More spots are ordered than will fit in the avails, or due to technical difficulties.

Bumper: A short announcement inserted between a program segment and the first commercial in an AVAIL. Especially used in children's programs. Examples**:** "Bugs Bunny will return after these messages." "And now back to the Big Movie." May also be called Transition.

Burn: Part of a pickup tubes that has a depletion of charge evidence by a negative image of the picture source causing the depletion. This condition is temporary unless the tube is turned off and allowed to cool before the problem is corrected.

Burned In: SMPTE\EBU time code numbers that are a permanent part of the corresponding video.

Burned-In Time Code: See BITC.

Burst Cutting Area: See BCA.

Burst Gate: This is a signal that tells a video decoder where the Colorburst is located within the scan line.

Burst: The part of the sync signal that controls the hue and color accuracy of television pictures.

Bus: A single row of buttons on a video switcher. Also a mixing network that combines the output of two or more channels.

Business: The realistic actions of actors that take place as a scene progresses. Examples would be reading a newspaper, knitting, lighting a cigarette, or fixing a drink.

Bust Shot: Camera shot of an actor from the waist up.

Busy: A scene that contains an inordinate amount of activity or distracting background elements.

Butt Splice: A film splice in which the film ends come together without overlapping. Also called a Butt-weld splice.

Butterfly Diffuser: A large net diffusion screen used between direct sunlight and a subject to soften the harshness of the sun.

Button Number: These are consecutive numbers assigned to each button on a menu, ranging from '1' to '36.'

Button: This is a rectangular area in the Sub-picture display area highlighted by the Highlight Information (HLI) that is used to define the active area on a menu associated with a specific action.

Buzz Track: A non-distinct background noise (sometimes called "room tone") usually recorded while everyone on the set remains silent. This recording can then be used be the editor to bridge any gaps between lines of dialogue.

BVB (Black-Video-Black): A preview mode used in conventional editing that displays black, newly inserted video, then black again. It allows the operator to see only the incoming material. It is not relevant to nonlinear editing systems as they usually offer far better ways of viewing material.

B-Wind: An indication of the emulsion position on a roll of film. A 'B' Wind indicates that the emulsion is facing out, or away, from the center or core of the film roll.

B-Y, R-Y The general term for color-difference video signals carrying blue and red color information, where the brightness (Y) has been subtracted from the blue and red RGB signals to create B-Y and R-Y color-difference signals.

Byte: A byte is composed of 8 bits. See Bit.

C

C Stand: A general purpose grip stand.

C: The color, or chrominance, signal representing all colors and already modulated onto a subcarrier as if to be combined with a luminance signal to become composite video. Its amplitude represents the color intensity (saturation) and its phase represents the color itself (hue). Unlike with the simplest and most common amplitude modulated signals, the sidebands of the C signal are not mirror images.

C-47: Ordinary wooden clothespins which are used to secure gels to barndoors. They are also known as a #1 wood clamp.

Cable Advertising Bureau (CAB): An organization of cable networks and cable systems that assists members in the promotion of cable television.

Cable Guards: Metal shields or "skirts" on a camera pedestal that sweep cables out of the way so the camera doesn't roll over them.

Cable Original: A program originally produced by a cable network or cable system.

Cable Penetration: The number of households or persons that can receive a cable network in their households either by a cable system, DBS, SMATV, etc. Nielsen updates cable penetration for its network subscribers every month. Penetration is based upon the installed People Meter sample.

Cable Ready: Television sets that have the circuitry built in them to receive and translate cable signals bypassing a cable converter box.

Cable Sync: A somewhat archaic method of sync sound shooting, where a cable runs from a Pilottone generator in the camera to the tape recorder.

Cable Television: Non-broadcast, point-to-point transmission of television programming by coaxial or fiber optic cable. Used for programming intended for a more limited audience than over-the-air broadcast television.

Cable: The electrical cords used to interconnect pieces of audio and video equipment.

Cache: Storage of digital data (video, audio, text, etc.).

CAD: Computer-aided design or Computer-assisted design.

California Scrim Set: A scrim set with two doubles.

Call Letters: A television station's three or four letter identification assigned to it by the FCC, and must be broadcast at least once per hour, at the Top of the Hour; e.g. KCSI-TV. Generally speaking, television stations east of the Mississippi River start with a W, while television stations west of the Mississippi River have call letters that start with a K. Canadian television stations generally start with a C, while in Mexico, the call letters generally begin with an X.

Call Sheet: A listing of which actors will be required for which scenes, and when they will be required. Call sheets are created by assistant directors and others.

Cam Head: A type of pan head for a camera mount. Commonly associated with studio cameras. Pan and tilt actions are controlled by variable drag adjustments.

Camcorder: Combination of camera and videotape recorder in one device. Camcorders permit easy and rapid photography and recording simultaneously. Camcorders are available in most home video formats: 8mm, Hi-8, VHS, VHS-C, S-VHS, etc.

Camel's-hair Brush: Brush made of soft camel's hair, often used for removing dust from lenses.

Cameo Background: A totally black background.

Cameo Lighting: The use of special lighting with a dark background so that subjects appear as if they are being photographed against a totally black background.

Cameo: A bit part played by a famous actor who would ordinarily not take such a small part. Originally meaning "a small piece of artwork," the term was borrowed by director Microphonehael Anderson when attempting to attract famous actors to play bit parts in Around the World in 80 Days.

Camera Angles: The viewpoints or angles from which the camera films the subject. The usual distinctions are High Angle, Straith-on, and Low Angle.

Camera Blocking: The process of notating the changing position of the camera, lens size, and focus during a particular scene.

Camera Cap: A cap screwed or mounted onto the front of a camera in place of the lens to protect the camera pickup tube from light or dirt when the lens is not in place.

Camera Car: A car or truck with a camera mount, used for the taping of scenes while the vehicle is in motion.

Camera Cards: Graphic cards designed to be picked up by a camera.

Camera Chain: A complete camera system: camera, cables, video controls and power supply.

Camera Control Unit (CCU): Controls for a camera where color correction, luminance, etc., are adjusted. CCUs for studio cameras are generally found in the video control room.

Camera Core: A 2 inch Core.

Camera Crew: The group of crewmembers directly involved with operation of the camera. Individual job titles include: clapper-loader, camera operator, assistant cameraman, director of photography, focus puller, grip, key grip, dolly grip, additional camera.

Camera Cue Light: A red light on the camera that indicates when it is or-the-air or being recorded. Also called a Tally Light.

Camera Head: The actual television camera which is at the "head" of a chain of camera-related electronics.

Camera Jib: Remotely-operated, crane-like camera mount which can typically move a camera from floor or ground level to a height of 10 or more feet.

Camera Left and Right: Directions given from the camera's point of view. The reverse of "stage left" and "stage right," used in theater.

Camera Lens: An optical system which concentrates, disperses, or changes the direction of light rays to form an image. The chief variable among lenses is the focal length. In technical terms, the focal length is the distance from the center of the lens to the point where the light rays converge to a point of focus. Focal length affects perspective in several ways: it alters the perceived magnification, depth, and scale of things in the image. We usually distinguish lenses on the basis of their effects on perspective. See Wide-angle Lens, Normal Lens, Telephoto Lens, Zoom Lens.

Camera Level: The relationship between the horizontal frame of the camera and the plane of the mise-en-scene.

Camera Light: A light normally mounted on the top of a camera to provide added illumination.

Camera Loader: The person who operates the clapboard at the beginning of a shot, also responsible for loading film stock into film magazines. The action of slapping the clapper was invented as a way of synchronizing the visual and audio components of a shot. Recent innovations in audio-visual synchronization have made this unnecessary, but it still occurs extensively. See also Assistant Cameraman.

Camera Log: A record sheet giving details of the scenes or shots photographed on a particular roll of negative.

Camera Movement: See Pan, Tilt, Tracking Shot, and Crane.

Camera Noise: The sound of the camera running. Even supposedly quiet cameras will make some noise.

Camera Obscura: A darkened chamber with a hole in one wall through which light enters, creating an image of the outdoors on the opposite wall. It was the earliest form of a "camera," and is where the name derives.

Camera of Record: Videotaping an event as it happens, generally without effects or production embellishments. Intended only as a "raw record" of the event.

Camera Operator: The person who operates the camera to the specifications dictated by the director of photography. A director or a director of photography sometimes assumes this role. (Luc Besson always operates the camera on films he directs.)

Camera Original: A slightly more adamant way of saying Original.

Camera Prompter: An electrical device for displaying a script that can be read by talent during a production. Also called Teleprompter.

Camera Rehearsal: Rehearsal using cameras, designed to make final checks before a production.

Camera Reports: A form of paperwork used to log shots and takes and put down any notes either to the lab or for future organization in the editing stage. There is generally one camera report per camera roll. Camera reports can be used to communicate specific timing requests to the lab (for instance, if a shot if lit with unusual color gels, this can be noted to let the timer know not to correct the color). Camera reports are extremely helpful to analyze any problem with the footage, since they provides a written record of the coverage (the least of which is that if the slate has the wrong information written on it, which happens now and then, a note can be made in the camera reports to keep the assistant editor from getting confused about which take is which).

Camera Roll: Each roll shot becomes a camera roll. It is often helpful to label them with a number in the order that they were shot. The usual way is with the abbreviation CR followed by a number. The lab will then assemble and print them in that order.

Camera Speed (Frame Rate): The rate at which a film travels through a camera. For 16mm and 35mm sound filming, the standard sound speed rate is 24 frames per second. There were no set speeds for silent films, but conventionally they are projected at 18 frames per second. The standard frame rate for motion picture camera projectors is 24 frames per second or 24fps. If you shoot motion picture film at a frame speed higher then 24fps you will get a slow motion effect. If you shot at a lower level such as 8 frames per second you will automatically speed up the action. When you shoot at different frame rates you must also remember to compensate for the exposure.

Camera Stock: This is film. It is also called camera stock to distinguish it from Print Stock.

Camera Supply: Most video cameras use an external DC voltage supply which is derived either from a battery belt worn by the camera operator, from a battery within the video recorder itself, or from the mains power supply (after voltage conversion).

Camera Tape: Cloth tape specifically for use on film shoots, much like gaffer's tape. Camera tape is typically 1 inch wide and white so that it can be used together with a sharpie for labeling magazines with the emulsion type and camera roll number. It is valid to use the terms gaffer's tape and camera tape interchangeably (they are both really the same type of tape) depending on how the tape is being used. It is designed not to leave a sticky residue behind on the camera.

Camera Tube: The primary light-sensitive glass tube in a tube-based video camera that converts the light image from the camera lens into an electrical signal.

Camera Wedges: Small wooden wedges, usually 4 inches long by 1/2 wide at the thickest end.

Cam-Lok: A feeder cable connector which is similar to a

Tweco, but with a more positive locking mechanism which is easier to disconnect.

Camp: A form of comedic parody where the clichéd conventions of a dramatic form like adventure are deliberately exaggerated to the point of ridiculousness. Often unfairly used to describe superhero films and shows as Batman is a prime example of this form of comedy.

Can: Metal or plastic containers used for the transport and storage of film.

Candela: A unit of light intensity, a standard candle.

Candlepower: The unit measure of incident light.

Canon Connector: .A three-pin connector used with balanced audio lines for line and microphone level audio signals. Also known as an XL or XLR Connector.

Canted Angle: See Dutch Tilt.

Canted Camera: Camera angle achieved by turning the camera slightly to one side causing the subject matter to run up or down hill. Result is also referred to as Dutch Angle.

Canted Frame: Often described as a Dutch Angle or Dutching. This is a device or process whereby the camera is angled so that the horizontal frame line is not parallel to the horizon.

Canvas Drop: A canvas background that can be lowered into place from the studio grid.

Cap Up: To cover the lens of a camera.

Capacitance Effect Disk: See CED.

Capacitance: An electrical component's ability to store electrical charges.

Capacitor: An electrical device which stores energy as an electrostatic charge. Often used as a component in filter circuits.

Capstan: A motor-driver roller and rotating shaft in the VTR that pulls the tape over the heads.

Captioning Service: Organization that encodes open or closed captions (text) into TV video—often for display and reading by hearing impaired.

Capture Card: Sometimes called a capture or video board, the logic card installed into a computer and used to digitize video. Or, for video that is already digitized, the device that simply transfers the file to the hard disk. Using a hardware or software codec, the capture card also compresses video in and decompresses video out for display on a television monitor.

Capture: Also called Digitization, the process of digitizing analog audio and video to binary files for editing, encoding, and digital distribution.

Cardioid Microphone: A Unidirectional microphone with most of its sensitivity aimed toward the front, and a pickup pattern that resembles an inverted heart.

Carrier Wave: An electromagnetic wave modulated with audio or video information used to broadcast radio and television signals.

Carrier: A frequency that is modulated with data to be transmitted.

Cart Cartridge: A plastic container with audio or videotape that threads automatically when inserted into a record or playback machine. In contrast to a cassette, the tape is commonly in a continuous loop.

Cart: Short for cartridge, a cart is the container holding magnetic tape, that typically has one or more Spots for an advertiser recorded on it. It is similar in size and operation to an 8-track tape. The cart is inserted into a playback machine for broadcast.

CAS: See Cinema Audio Society.

Cascading: The flow of a signal through a sequence of electronic stages, each stage adding an additional effect.

Cash Syndication: A transaction whereby the television station pays the syndicator a cash license for the right to air a show over a period of time (i.e., six runs over four years). In a cash agreement, the television station bears all the risk; the syndicator gets all the money regardless of how well the program performs. All commercial inventory is sold by the television station, hence no national spots. Many popular off-network shows (i.e., Murphy Brown and Cheers) are sold on a straight cash basis.

Cash/barter Syndication: A combination of cash syndication and barter syndication in which the television station gives up commercial time in addition to paying a license fee for the rights to air a program. Typically the television station has more commercial inventory to sell than a straight barter agreement and has a lower fee than a cash only agreement.

Cassette: A plastic container with a supply and take-up reel of audio or videotape that threads automatically when placed in a record or playback machine. Unlike the cartridge, the tape is not in a continuous loop and must be rewound.

Cast: A collective term for the actors appearing in a particular movie.

Casting Couch: Slang term referring to "Golden Age" of Hollywood, where wannabe stars would grant sexual favors to directors/producers in return for a role in their film.

Casting Director: The person who auditions and helps to select all of the speaking role actors in film, television shows or plays. The CD must possess a vast knowledge of the actor pool and be able to match a variety of actors with just the right role. Directors and producers rely on the Casting Director to assist them with assembling the perfect cast for their production. Casting Directors are also responsible for serving as the liaison between the director, and the actors and their agents. CDs negotiate the deals with agents once the actors have been cast and are also responsible for the contracts and SAG of each actor.

Casting Society of America (CSA): A professional organization of Casting Directors working in theatre, film, and television. The CSA is not a union or a guild, therefore every Casting Director working in these mediums is not necessarily a member of this organization.

Casting: The process of hiring actors to play the characters in a script, typically done by a casting director, but with some input from a director, producer, or studio.

Castle Rock Broadcast Center: The facility which provides all DirecTV television reception, tape playback, encoding, and uplinking.

Catadiatropic Lens: A telescopic lens that uses an internal mirror to reduce the physical length of the lens. Also called a Cat Lens.

Catchlight: A small, concentrated reflection from a major light source which appears in a subject's eyes.

Cathode Ray Tube (CRT): A television picture tube. The cathode ray excites the pixels to create the video image.

CATV: Acronym for cable TV, derived from the older term, community antenna television.

Causality: The arrangement of scenes in editing to suggest cause and effect relationships.

Cause-effect Chain: In narrative structure, the way one event leads to (causes) another and is the result (effect) of a previous event.

CAV: See Constant Angular Velocity.

Cb, Cr: Refers to the color component video signals B-Y, R-Y respectively optmized for digital purposes or transmission. Sometimes loosely used to refer to B-Y (Pb) and R-Y (Pr) in any context.

C-Band: The range of frequencies from four to six gigahertz (billions of cycles per second) used by most communications satellites.

CBC: Canadian Broadcasting Corporation.

CBEMA: Computer and Business Equipment Manufacturers Association.

CBR (Constant Bit Rate): A variety of MPEG video compression where the amount of compression does not change.

CC Filters: Color compensating filters made in precise density values of the primary and secondary colors.

CCD: See Charge Coupled Device.

CCI (Copy Control Information): Information specifying if content is allowed to be copied.

CCIR: See International Radio Consultative Committee.

CCIR-601: The international standard for conversion of analog video to digital video, now identified as ITU-R-601. ITU-R-601 specifies 720 samples per line, yielding a frame size of 720x480 (NTSC) or 720x576 (PAL). ITU-R-601 is the international standard for professional video applications.

C-clamp: A screw-down clamp used to mount lights to studio grid pipes and lighting stands.

CCTV (Closed Circuit TV): A video system used in many commercial installations for specific purposes such as security, medical and educational.

CCU: See Camera Control Unit.

CD (Compact Disc): An audio recording and playback medium that uses digitally encoded information on a laserdisc. The sound is decoded by optical laser for high-quality audio playback.

CD-i (Compact Disc Interactive): An extension of the CD format designed around a set-top computer that connects to a TV to provide interactive home entertainment, including digital audio and video, video games, and software applications. Defined by the Green Book standard.

CD-R (Compact Disk, Recordable): Refers to compact disks that can be recorded one at a time, or the equipment used to make the recordings. The CD-R disk can only be recorded once; if the recorded content is no longer wanted, additional material can be recorded only on the space remaining subject to the recording method chosen, or if there is no remaining space the disk is discarded. The disks themselves are constructed differently from ordinary non-recordable CD's. Ordinary CD's are stamped like grooved records, and then the silver or gold reflective layer applied. Recordable CD's have the foil layer (gold colored) already applied, and the recording consists of "burning" the pits into an intervening layer of organic dye one at a time. Current technology allows recordings to be made in 1/4 the time that an audio CD plays (1/4 of 74 minutes or 19 minutes per side) although owners of consumer grade CD recorders have tended to find that recording faster than "2x" often yields a defective disk.

CD-RW (Compact Disk, Rewritable): Refers to compact disks that can be recorded one a time and where an unwanted recording can be recorded over. The disks themselves are constructed differently from other types of CD's; they contain an organic dye that represents the pits that in turn represent the recorded content. "The jury is out" as far as the longevity of these disks goes; to this writer the organic content suggests possible aging problems that color film today still has. Not all CD players can play CD-RW disks since the reflective surface has a different color from the usual silver or gold.

CDV: A combination of laserdisc and CD which places a section of CD-format audio on the beginning of the disc and a section of laserdisc-format video on the remainder of the disc.

CEA: See Consumer Electronics Association.

CED (Capacitance Effect Disk): Also referred to as "RCA Selectavision" video disks. This is a 12 inch video disk and player system marketed by RCA, and now obsolete. It did not have consumer recording capability. No laser was used; the disks were grooved like (pre-CD) phonograph records, and the video signal was recorded as rising and falling ripples ("hill and dale" in older phonograph terminology). The needle, or stylus, does not follow the ripples exactly, it is not small enough or given enough pressure to.

Cell Animation: Also called onion skinning, an animation technique in which a background painting is held in place while a series of transparent sheets of celluloid containing objects are placed over the background painting, producing the illusion of movement. One of the two main types of animation associated with digital video. Compare frame-based 2-D animation.

Cell Command: A Navigation command executed when the presentation of a cell has been completed.

Cell: In DVD-Video, a unit of video anywhere from a fraction of a second to hours long. Cells allow the video to be grouped for sharing content among titles, interleaving for multiple angles, and so on.

Celluloid: A transparent sheet of plastic, on which images are drawn and painted in the production of animation. Sometimes referred to as Cell.

Celo: A type of cookie which is made from wire mesh coated with plastic.

CEMA: See Consumer Electronics Manufacturers Association.

Cement Splice: A type of splice used primarily by negative cutters. In a cement splice the two pieces of film overlap each other and are fused together with film cement.

Censorship: Changes required of a movie by some person or body other than the studio or the filmmakers, usually a national or regional film classification board. See also Certificate.

Center Marker: A cross that indicates the center of the image on the viewfinder screen.

Center Spot Filter: A diffusion filter with a clear central area. Gives a clear central area and slightly diffuses surrounding areas.

Center Track: A standard position for the placement of the audio signal on magnetic film. The placement is in a narrow band centered between the two edges of the film.

Centre International du Film pour l'Enfance et la Jeunesse (CIFEJ): A 40-year-old international non-governmental organization whose goal is to promote quality films, television programs and videos for children and young people around the world.

Century Stand (C-Stand): A multipurpose grip stand.

Certificate of Authorship: Form signed by the author of a screenplay or other written work that warrants the author's work is original, does not libel another party, does not invade anyone's privacy, and will not cause the buyer of the work to be sued for any legal action.

CG: See Character Generator.

CGA (Color Graphics Adapter): A relatively low-resolution color display system for PC-type computers.

Computer-generated Imagery (CGI): An image that is synthetic and can appear to be either graphic or photo-realistic in nature. A CGI image is one that originates in the computer.

CGI: See Computer-generated Imagery.

CGMS (Copy Generation Management System): Copyright protection method of making it impossible for most people to copy a DVD.

Chain Vise Grip: A vise grip with a chain that is used for its clamping capability.

Challenge Key: Data used in the authentication key exchange process between a DVD-ROM drive and a host computer, where one side determines if the other side contains the necessary authorized keys and algorithms for passing encrypted (scrambled) data.

Change Pages: When a script is being edited during production, changes are distributed to actors and the filmmakers on "change pages," which are usually a different color to the pages of the script.

Changeover Cue: A mark at the end of a release print reel to indicate the moment at which to switch over to the next reel on the second projector.

Changing Bag: A double chambered black bag with a zipper on one end and two elasticized arm holes on the other side, used for loading film into magazines.

Channel Bit: The bits stored on the disc, after being modulated.

Channel Capacity: The number of channels available for use on a current (or future) cable system.

Channel Coding: Describes the way in which the IS and OS of the data stream are represented on the transmission path.

Channel Data: The bits physically recorded on an optical disc after error-correction encoding and modulation. Because of the extra information and processing, channel data is larger than the user data contained within it.

Channel Mapping: The placement on the dial of a cable network or television station on a particular cable system. Usually, the lower a cable network appears on a cable system (i.e., channels 2 to 13) the higher the ratings will be for the cable network.

Channel: A band or frequency assigned to a broadcast television station.

Chaoji VideoCD: Another name for Super VideoCD.

Chapter: In DVD-Video, a division of a title. Technically called a part of title (PTT).

Character Actor: An actor who specializes in playing a particular style of character, often stereotypical, offbeat, or humorous.

Character Arc: Formulaic inferred curved line which traces the development, growth, and transformation of a character over the course of the screenplay.

Character Generator (CG): A small computer used to generate titles and other text electronically without the use of a camera.

Character Make-up: Make-up—generally accompanied by hair pieces, etc.—designed to change an actor's appearance into that of the character being played.

Charge Back: Charging studio costs to another division of the same company.

Charge Coupled Device (CCD): A semiconductor device (IC) that converts optical images to electronic signals. CCDs are the most commonly found type of image sensor in consumer camcorders and video cameras.

Charger: An electrical, AC-to-DC device which replenishes (recharges) battery power.

Cheat: To arrange talent or props so that an optimum camera angle is achieved, often without regard to how realistic it may be for actors. When the camera is set up for a second shot at a different angle it is possible to move things around a little to improve the new composition, the difference in perspective and angle of the two shots hiding the fact that things are not exactly in the same place. Both actors and furniture on the set can be cheated. The term is often used as cheating something "into" a shot or "out of" a shot, as in telling an actor "We're going to cheat you in a little," and having them stand a little to one side so more of them is in the shot.

Check Print: This is a print made from an internegative or an optical to verify the quality and success of an effect.

Checkerboard Cutting: A technique of splicing A/B rolls in order to eliminate the image of the film splice from the duplicates or to allow for rapid readjustments of optical printer settings from shot to shot.

Checkerboard Programming: A programming tactic where the television station or network airs different programs each night in a particular time period to try to appeal to a wider audience and build higher ratings.

Typically occurs in prime access with different syndicated programs.

Checksum: An error-detecting scheme which is the sum of the data values transmitted. The receiver computes the sum of the received data values and compares it to the transmitted sum. If they are equal, the transmission was error-free.

Chiaroscuro: A low-key lighting style, usually in reference to theatrical productions or the dark paintings of Rembrandt.

Chip Camera: A camera that uses one or more CCD's. See CCD.

Chip Chart: A test chart consisting of shades of gray from TV white to TV black used for setting up cameras.

Choreographer: A person who plans and directs dance sequences within a movie.

Chroma Bandpass: In a NTSC or PAL video signal, the luma (black and white) and the chroma (color) information are combined together. If you want to decode an NTSC or PAL video signal, the luma and chroma must be separated. A chroma bandpass filter removes the luma from the video signal, leaving the chroma relatively intact. This works reasonably well except in images where the luma and chroma information overlap.

Chroma Burst: See Colorburst.

Chroma Channels: The separate light-sensitive pickup devices or color circuits in a TV camera. Typically there are three channels: red, green, and blue.

Chroma Control: The process of controlling the color and saturation information in a video signal.

Chroma Corrector: A device used to correct problems related to the chroma of the video signal, as well as color balance and color noise.

Chroma Crawl: An undesirable moving color artifact or

pattern in an NTSC picture. Generally caused by a television set decoding color information as high-detailed luminance information.

Chroma Demodulator: After the NTSC or PAL video signal makes its way through the Y/C separator, the colors must be decoded. That's what a chroma demodulator does. It takes the chroma output of the Y/C separator and recovers two color difference signals (typically I and Q or U and V). Now, with the luma information and two color difference signals, the video system can figure out what colors to display.

Chroma Gain: Camera control that controls the amounts of color in a picture.

Chroma Key Blue: A deep shade of blue commonly used in backgrounds which is intended to be keyed out and replaced with other video. See Chroma Key.

Chroma Key: Process of overlaying one video signal over another by replacing a range of colors with the second signal. An example of chroma keying in action is the nightly news person standing in front of a giant weather map. In actuality, the person is standing in front of a blue or green background and their image is mixed with a computer-generated weather map. This is how it works: a TV camera is pointed at the person and fed along with the image of the weather map into a box. Inside the box, a decision is made. Wherever it sees the blue or green background, it displays the weather map. Otherwise, it shows the person. So, whenever the person moves around, the box figures out where he is, and displays the appropriate image.

Chroma Noise: Noise which manifests itself in a video picture as colored snow.

Chroma Processor: A computer program associated with desktop editing programs that can subtly or drastically alter colors and color balance in sources of video.

Chroma Resolution: As distinct from luminance detail, the amount of detail in color channels.

Chroma Subsampling: Reducing color resolution by taking fewer color samples than luminance samples.

Chroma Trap: In a NTSC or PAL video signal, the luma (black and white) and the chroma (color) information are combined together. If you want to decode the video signal, the luma and chroma must be separated. The chroma trap is one method for separating the chroma from the luma, leaving the luma relatively intact.

Chroma: Color information in a video signal, consisting of hue (phase angle) and saturation (amplitude) of the color subcarrier signal.

Chromatic Aberration: Color fringing or other color abnormalities caused by imperfections in a camera lens.

Chromatic Sensitivity: See Monochromatic Stock, Orthochromatic Stock, and Panchromatic stock.

Chrominance Color: TV camera color channel.

Chrominance Signal: A video signal containing color information.

Chrominance: The color portion of a video signal separate from the luminance component, representing the saturation and tint at a particular point of the image. Black, gray, and white have no chrominance, but any colored signal has both chrominance and luminance. The higher the chrominance level, the stronger the color (e.g., a strong signal produces red, and a weak signal, pink).

Churn: A cable industry measure based on a formula that measures subscriber connects, disconnects, upgrades and downgrades in service. A low churn rate typically indicates customer satisfaction.

CIF (Common Interface Format): The Common Interface Format was developed to support video conferencing. It has an active resolution of 352 x 288 and a refresh rate of 29.97 frames per second. The High-Definition Common Image Format (HD-CIF) is used for HDTV production and distribution, having an active resolution of 1920 x 1080 with a frame refresh rate of 23.976, 24, 29.97, 30, 50, 59.94, or 60 Hz. Sometimes referred to as Common Image Format.

CIFEJ: See Centre International du Film pour l'Enfance et la Jeunesse.

Cinch Marks: Not to be confused with sync marks. Cinch marks are small vertical scratches on a roll of film that are caused when the end of the film is pulled to tighten the roll, causing any dust on the film to make a small scratch. Too much drag on the supply while rewinding is one common way that cinching can occur.

Cinema Audio Society (CAS): A philanthropic, non-profit organization formed in 1964 for the purpose of sharing information with sound professionals in the motion picture and television industry. Use of the abbreviation after a name indicates that the person is a member of the CAS.

Cinema Verité: Literally: Cinema Truth. A documentary style in which no directorial control is exerted. The term is frequently misused to describe new-wave "handheld" camera techniques.

Cinema: A place where screenings occur. Cinemas can be hardtops or ozoners.

Cinemascope: The trade name for wide-screen films photographed and projected with anamorphic lenses on the camera and the projector.

Cinematographer: The person that determines the exact nature of the lighting, camera placement, exposure and movement that is performed during the filming of live or motion controlled photographic elements in a media project.

Cinematography: The process of making a film image, and the characteristics of the film image.

Cinepak: A commonly used QuickTime codec for compression of video files on CD-ROM. Cinipak offers temporal and spatial compression, and data-rate limiting.

Cinerama: A wide-screen motion picture process that employs three cameras and three projectors, a wide curved screen, and stereophonic sound. Three separate images are projected simultaneously onto the curved screen, widening the picture into the viewer's peripheral vision.

Cinex Strip: A short test print in which each frame has been printed at a different exposure level.

CIRC (Cross-interleaved Reed Solomon Code): An error-correction coding method which overlaps small frames of data.

Circuit Breaker: An electronic "fuse" that can be reset after it is overloaded.

Clamp Light: A type of lighting fixture designed to hold a screw-in light bulb, with a not-so-dependable spring clamp for mounting on the side of an open door, etc. Often includes an aluminum reflector dish as well.

Clamp: This is basically another name for the DC-restoration circuit. It can also refer to a switch used within the DC-restoration circuit. When it means DC restoration, then it's usually used as "clamping." When it's the switch, then it's just "clamp."

Clamping Area: The area near the inner hole of a disc where the drive grips the disc in order to spin it.

Clapboard: Small board which holds information identifying a shot: working title of the movie, names of director and DP, scene and take numbers, date, and time. Used at the beginning of a take, the clapboard has a hinged stick which is "clapped" to provide audio/visual synchronization.

Clapper: The sticks that are slapped together in view of the camera for the purpose of synchronizing film sound. These are usually, but not always, attached to the slate and appear at the head or tail of a sync sound take.

Clapper-Loader: Individual who operates the clapboard at the beginning of a shot, also responsible for loading film magazines into the camera.

Clarke Belt: Named after its founder Arthur C. Clarke, the Clarke Belt is an orbit used by satellites at a height of 22,250 miles, in which satellites make an orbit in 24 hours, yet remain in a fixed position relative to the earth's surface.

Classical Hollywood Cinema: A conventional style of filmmaking with a particular model of narrative structure, editing technique, mise-en-scene, dialogue, music, etc. Narrative is presented in a clear cause-and-effect chain, with definite closure. Also called Hollywood Classicism.

Classical Period: In the history of theatrical cinema, the 1920s-50s when the Hollywood studio system of film production held total power and evolved the classical style of filmmaking. In a genre's evolutionary pattern, the stage during which thematics, narrative structure and audial/ visual style are solidified into firm conventions, a recognizable cohesive unit.

Claw: Used to hold film in place during exposure.

Claymation: Animation of models constructed from clay or plasticine.

Clean List: An edit decision list (EDL) used for linear editing that has no redundant or overlapping edits. Changes made during off-line editing often result in edits that overlap or become redundant. Most computer-based editing systems can clean an EDL automatically. Contrast with dirty list (dirty EDL). See also On-line editing.

Clean Speech: A take in which all the dialogue was performed without error.

Clear: Non-scrambled satellite TV programming.

Clearance: To obtain a time period for a program on a television station or cable system.

Click Track: A prerecorded track of electronic metronomic clicks used to ensure proper timing of music to be recorded. Essential in music scoring sessions.

Cliff Edge Effect: An RF (radio frequency) characteristic that causes DTV reception to deteriorate dramatically with a small change in signal reception. Bit error rate increases to the point where video cannot be obtained by the receiver. The picture and audio are lost entirely.

Cliff Effect: When approaching the fringes of reception, analog TV pictures begin to degrade by becoming "snowy." By contrast, when in a weak digital signal area, a relatively small change in received power may cause the DTV picture to abruptly change from perfect to nothing; hence the name, "cliff effect."

Cliffhanger: A moment of high drama, frequently used at the end of serials. Named for the (now clichéd) practice of leaving a hero or heroine hanging onto the edge of a cliff.

Clip Art: Professionally made drawings, generally available on CDs, that can be used in newsletters and various media.

Clip Audio: The Quantel term for audio that has been recorded with, or added to, a video clip. An audio-only clip is stored with only one frame of video repeated for the audio duration.

Clip Bin: Editing screen display in a window that shows video clips that have been digitized and ready of use.

Clip Pack: A multi-layer clip constructed in the Pack Reel of the Edit Desk in Quantel editing systems.

Clip Properties: A clip's specific settings, including frame size, compressor, audio rate, etc.

Clip Service: A service that distributes segments of audio and video content to Web sites for sales and demo purposes.

Clip: The name is taken from the film industry and refers to a segment of sequential frames made during the filming of a scene. In broadcast terms a clip is the same, but represents a segment of video frames.

Cliplink: Sony DVCAM approach for marking in/out points of raw footage while it's in the camera. Resulting thumbnail images and time code numbers can then be downloaded to the nonlinear editor as a guide in digitizing only the selected shots.

Clip-on Microphone: A small personal microphone that can be attached to clothing.

Clipper Gain Knob: A knob on a video effects unit used to set the threshold level for color or luminance keying.

Clipping Logic: A circuit used to prevent illegal conversion. Some colors can exist in one color space but not in another. Right after the conversion from one color space to another, a color space converter might check for illegal colors. If any appear, the clipping logic is used to limit, or clip, part of the information until a legal color can be represented. Since this circuit clips off some information and is built using logic, it's not too hard to see how the name "clipping logic" was developed.

Clipping: Cutting of the peaks of a signal. In video this generally affects the while level.

Clock Jitter: Timing uncertainty of the data cell edges in a digital signal.

Clock Recovery: The reconstruction of timing information from digital data.

Clock-In: Probably the universal method of converting an analog video signal to digital. Since an incoming video signal arrives one pixel at a time, it is sampled at fixed time intervals (in nanoseconds). Essentially the signal is chopped up into little pieces, much as a chef may chop carrots or celery or bananas into slices. Clocking in has the shortcoming of sometimes taking the last half of one pixel and the first half of the next pixel (pixel straddling) as a new pixel when the incoming analog video was digital at some earlier time; the former pixel footprint remains. This results in loss of horizontal resolution. Interpolation may occur naturally during clock-in and the result varies depending on whether the digital value (sample) is derived from the entire slice (averaged) or only a small portion of the slice.

Clone: A copy of a non-compressed digital tape to another non-compressed digital tape. As the copy will be indistinguishable from the original it is referred to as a clone.

Close Captioned (CC): System which displays the current dialog on screen for deaf or hard-of-hearing viewers.

Close Focusing: Zoom lenses often focus no closer than 3 feet if they are in front of an object or scene. Prime lenses will focus if the camera is closer and macro lenses will usually focus at a ratio of 1:1.

Close Microphone: The positioning of a microphone very close to the performer's mouth—often used by radio and TV announcers.

Close: See Billboard.

Closed Circuit: A method of program transmission that is limited to a fixed audience, typically by special microwave or cable equipment.

Closed Set: A television studio that is closed to all but specific crew and talent involved in the scene.

Closed Subtitles: See Subtitles.

Close-up (CU): A framing that presents a close view of an object or person: filling the frame and separating it or her or him from the surroundings. Conventionally, a TV close-up of a person is from the shoulders or neck up.

Closure: Occurs when enigmas opened at the beginning of a program and throughout are resolved; all of the narrative's questions are answered. The opposite of aperture.

CLUT: See Color Lookup Table.

Clutter: All of the non-program elements within a given program or time period, such as commercials, PSAs, promos, IDs, billboards, etc. Television stations try hard to increase revenues while trying to keep clutter to a minimum.

CLV: See Constant Linear Velocity.

CMI: See Content Management Information.

C-mode: A method of editing where the EDL is arranged by source tape reel numbers with the required material from each listed in advancing time code: giving a quick, convenient way of accessing the material, minimizing both spooling and tape changes. This means the edit may well have to be assembled out of final program sequence.

C-mount Lens: A lens with a standard one-inch threaded mount assembly that is screwed into the camera body, as opposed to a bayonet or "m" mount.

CMYK: This is a color space primarily used in color printing. CMYK is an acronym for Cyan, Magenta, Yellow, and blacK. The CMYK color space is subtractive, meaning that cyan, magenta, yellow and black pigments or inks are applied to a white surface to remove color information from the white surface to create the final color. The reason black is used is because even if a printer could print hues of cyan, magenta, and yellow inks perfectly enough to make black (which it can't for large areas), it would be too expensive since colored inks cost more than black inks. So, when black is used, instead of putting down a lot of CMY, they just use black.

Coated Lens: Lenses and lens elements having a magnesium-fluoride coating designed to reduce light reflections.

Coaxial Cable: The standard cable consisting of a central inner conductor and a cylindrical outer conductor. Used for many video connections, especially by CATV companies.

Coaxial Magazine: A type of magazine with two chambers side by side, with the supply and take up rolls rather like wheels mounted on either end of the same axle.

Co-channel Interference: Interference caused by two or more television stations using the same channel (frequency).

Code Numbers: Inked-on edge numbers, usually added to a workprint and mag track after syncing, so that corresponding sound and picture can always be properly aligned during editing. They are also used for the general organization of the footage. Sometimes the term edge numbers are used, and although this is not incorrect, care should be taken that it is understood that you are talking about the inked-on numbers and not the Latent Edge Numbers.

Code: A set of rules.

CODEC (Coder -Decoder): A device that converts analog video and audio signals into a digital format for transmission over telecommunications facilities and also converts received digital signals back into analog format.

Coded Orthogonal Frequency Division Multiplexing (COFDM): COFDM can transmit many streams of data simultaneously, each one occupying only a small portion of the total available bandwidth. It's the DTV standard used in Europe.

Coding: Representing an information signal as a series of numbers, usually in binary form. Once the workprint and sound stock (mag) have been placed in sync, the rolls are coded with matching yellow edge numbers so they can be matched up later once they have been cut up into pieces.

COFDM: See Coded Orthogonal Frequency Division Multiplexing.

Coherence: A descriptive dimension of light that characterizes its hardness or softness. The harder a light source is, the more coherence it is said to have. Professional photographers refer to coherence as light quality.

Coincidental Ratings: Ratings based on information obtained from viewers or listeners at the time of the broadcast, usually through telephone interviews.

Color Announcer: A type of television sports announcer; often he or she is a former athlete and/or coach, with first-hand expertise.

Color Balance: Adjusting the chroma channels in a video camera so that whites in a scene will be reproduced as true white.

Color Bar Generator: Device that create color bar test signal.

Color Bar Signal: A test signal which can be displayed as vertical bars of different colors on a color video monitor. It is used to check chrominance functions of color television and camera's.

Color Bars: An electronically generated video pattern consisting of eight equal width colors, used to establish a proper color reference before recording and playback and for adjustment purposes.

Color Compatibility: The ability of different color subject matter to translate into a black and white picture and maintain adequate tonal contrast and separation.

Color Consultant: A technical advisor with expertise in film stock and film developing, who provides advice for cinematographers and color timers.

Color Conversion Filter: A camera filter used to convert light from indoor to outdoor color temperature.

Color Correction: The electronic altering of the coloring of a video image. See also Chroma Corrector.

Color Decoder: A device which divides a video signal into its basic color components. In TV and video, color decoding is used to derive signals required by a video monitor from the composite or Y/C signals.

Color Demodulator: See Chroma Demodulator.

Color Depth: The number of levels of color (usually including luma and chroma) that can be represented by a pixel. Generally expressed as a number of bits or a number of colors. The color depth of MPEG video in DVD is 24 bits, although the chroma component is shared across 4 pixels (averaging 12 actual bits per pixel).

Color Difference: All of the color spaces used in color video require three components. These might be R'G'B', Y'IQ, Y'UV or Y'(R': Y')(B': Y'). In the Y'(R': Y')(B': Y') color space, the R': Y' and B': Y' components are often referred to as color difference signals for obvious reasons. They are made by subtracting the luma (Y') from the red and blue components. I and Q and U and V are also color difference signals since they are scaled versions of R'**:** Y' and B'**:** Y.' All the Ys in each of the Y'IQ, Y'UV and Y'(R'**:** Y')(B'**:** Y') are basically the same, although they are slightly different between SDTV and HDTV.

Color Edging: Extraneous colors that appear along the edges of objects, but don't have a color relationship to those areas.

Color Encoder: The color encoder does the exact opposite of the color decoder. It takes two color difference signals, such as I and Q or U and V, and combines them into a chroma signal. In broadcast television, the transformation of primary color signals (or luminance and color difference signals) into composite video.

Color Frame: A sequence of four fields (2 frames) of NTSC video. Since the phase of the color subcarrier reverses every frame in NTSC (which makes the color information essentially invisible on a black-and-white receiver) two complete frames (four fields) are required to carry a complete sequence of color information. A PAL color frame consists of four frames (eight fields).

Color Key: This is essentially the same thing as Chroma Key.

Color Killer: A circuit that shuts off the color decoding if the incoming video does not contain color information. The color killer looks for the colorburst and if it can't find it, it shuts off the color decoding.

Color LCD: A flat, liquid-quartz-type display which can reproduce a full color image.

Color Lookup Table (CLUT): An index that maps a limited range color values to a full range of values such as RGB or YUV.

Color Media: See Gel.

Color Model: Any of several means of specifying colors according to their individual components. See RGB, YUV.

Color Modulator: See Color Encoder.

Color Phase: The phase of the chroma signal as compared to the colorburst, is one of the factors that determines a video signal's color balance.

Color Process: See Tinting and Toning.

Color Processing: A way to alter a video signal to affect the colors. See also Chroma Corrector.

Color Purity: This term is used to describe how close a color is to the theoretical. For example, in the Y'UV color space, color purity is specified as a percentage of saturation and +/-q, where q is an angle in degrees, and both quantities are referenced to the color of interest. The smaller the numbers, the closer the actual color is to the color that it's really supposed to be. For a studio-grade device, the saturation is +/-2% and the hue is +/-2 degrees. On a vectorscope, if it's in that range, it's studio quality.

Color Resolution: The degree to which the combined color channels in a system can reproduce fine details.

Color Reversal Intermediate (CRI): This is a duplicate color negative which is made by the reversal process.

Color Space Conversion: The transformation of video from one color space representation (such as YUV) to another color space (such as RGB). Color space conversion is normally required during video rendering. This conversion can be performed in software or hardware, however software conversion results in longer render times.

Color Space: A mathematical representation for a color.

Color Standard: A color criterion or reference used to setup color video equipment.

Color Subcarrier: The 3.58 MHz (NTSC) or 4.43 MHz (PAL) signal that carries color information. The color subcarrier is superimposed onto the luminance level. Its amplitude represents saturation and its phase angle represents hue.

Color Temperature Blue: See CTB.

Color Temperature Orange: See CTO.

Color Temperature: A method for specifying the overall color of a light source, measured in degrees Kelvin (deg.K). If a TV has a color temperature of 8,000 degrees Kelvin, that means the whites have the same shade as a piece of pure carbon heated to that temperature. Low color temperatures have a shift towards red; high color temperatures have a shift towards blue.

Color Timing: The process wherein colors are referenced and alternate odd and even color fields are matched to ensure colors match from shot to shot. Most commonly found in high-end equipment, such as Betacam SP.

Color Under: Heterodyne-type color process used in most videocassette recorders. To facilitate recording (with some loss of quality) the basic 3.58MHz color frequency is converted to a lower frequency.

Colorburst: The portion of the composite video signal that contains a sample of the color subcarrier. The colorburst is used to establish a reference for the color information that follows it, and for decoding the color information of the signal. Burst is many cycles of 3.58 MHz pulses recorded during the horizontal blanking interval and used to establish phase relationships for determining the hue. A color television receiver's color oscillator is phase locked to the colorburst.

Colored Filter: A glass or plastic material placed either in back or in front of a camera lens to subtract certain colors of light. Used to alter the colors in a scene.

Colorimetry: Characteristics of color reproduction. In TV, the range of colors that a system can reproduce.

Colorist: The timer of a video transfer.

Colorization: Film alteration process where an operator

digitally alters a black and white image to include color.

Colorizer: An electronic circuit that generates color which is added to or made to modify an image.

Colorizing Control: A knob which can move through the entire range of colors. Used to select colors for electronic effects.

Colorstream: The name Toshiba uses for the analog YPbPr video interface on their consumer equipment. If the interface supports progressive SDTV resolutions, it is called ColorStream Pro. If the interface supports HDTV resolutions, it is called ColorStream HD.

Color-subcarrier: A sub-channel of a broadcast signal designed to add color information to a basic black-and-white video signal.

Comb Filter: Device using the periodic repetition of chrominance subcarrier phases along the vertical axis and sometimes the temporal axis to perform luminance-chrominance separation in the decoder. In the signal frequency domain, its response has a periodic (comb-like) shape. A comb filter is used in place of a chroma bandpass or chroma trap. The comb filter provides better video quality since it does a better job of separating the luma from chroma. It's called a comb filter because the frequency response looks like a comb.

Combo Box: A six pocket stage box that can be converted from three-phase four-wire to single phase three-wire operation.

Combo Stand: A heavy duty 2K stand without wheels. It is called a combo because it can be used for both reflectors and lights.

Combo: An abbreviation for "combined television stations," most commonly referring to a method of selling commercials for either an AM/FM situation, or any combination of TV pilot and satellite television stations.

Comet Tail: A smear or black trailing image in video

cameras used under certain conditions. Caused by subject matter in motion.

Commercial Appropriation: A legal aspect of invasion of privacy relating to the unauthorized used of the name or likeness of a well-known individual for the personal or economic gain of a third party.

Commercial Protection: A broadcaster's guarantee to an advertiser stating that his commercial will have a given amount of separation from direct competitors. Also called Competitive Separation.

Common Interface Format: See CIF

Communication Protocol: A specific software based protocol or language for linking several devices together. Communication protocols are used between computers and VCRs or edit controllers to allow bi-directional "conversation" between the units. See also RS-232/RS-422.

Communications Satellite: An electronic retransmission vehicle located in space in a fixed geosynchronous orbit 22,300 miles above the earth's equator. Communications satellites are used by broadcast networks, syndicators, and cable networks.

Compander: Refers to a device that compresses an input signal and expands the output signal in order to reduce noise.

Comparative Renewal: The process in which the FCC decides whether to renew a broadcaster's television station license upon its expiration or to award it to another rival applicant.

Compatibility: The ability of the elements in a color television picture to be reproduced on a black and white receiver with adequate tonal contrast and subject matter separation.

Compensation: With regard to writers, the money paid for writing services or for the sale of a screenplay.

Competitives: Requests from advertisers or agencies for

information (rates, program changes, etc) on all competing television stations in the market, usually pertaining to specific products.

Compilation Editing: Commonly used editing style in documentaries. Continuity can jump around and not adhere to any logical or time-based sequence. In compilation editing continuity is usually provided by narration.

Compilation Film: A film composed largely of stock footage or clips from other films.

Compiler: A special computer program that translates a series of instructions written by a computer programmer in a high-level programming language into a machine-level instruction that the system can execute directly.

Complementary Colors: Colors that are opposite each other on the color wheel.

Completion Bond: An insurance guarantee that principal photography on a given film will be completed. It indemnifies a production against the unforeseen costs of any type, whether or not they result from problems which are covered by other types of insurance.

Complexity Editing: Assembling shots with a primary regard to building story or dramatic effect rather than following a logical or temporal sequence.

Component Digital: A digital representation of a component analog signal set, most often Y, B-Y, R-Y.

Component Recording: Videotape recording in which the color information is recorded separately from the luminance and synchronization information.

Component Signals: In contrast to a composite signal, non-encoded video signals. Typically, refers to separate luminance and chrominance signals.

Component Television: Home television systems that consist of a separate video monitor, tuner, audio amplifier and speakers.

Component Video: A recording method that does not require NTSC encoding of RGB signals, but records and routes luminance and color signals separately. Because component video neither reduces bandwidth nor compromises the RGB components, it produces images of higher resolution and better color than composite video. Although component video was originally designed for high-end television production, similar recording methods are now available with consumer-level equipment.

Component: Refers to the use of independent, individually controlled audio or video elements in a single television or audio system. A TV system having a separate monitor, tuner, and audio system would be an example.

Composer: A musician whose music appears in a movie's score. Most movies have at least some original music written for the score, usually after the relevant parts of the movie have been filmed. See also Lyrics.

Composite Master: Final, end product of an editing session including all program elements and special effects.

Composite Print: A motion picture print with both picture and sound on the same strip of film.

Composite Sync: A signal consisting of horizontal sync pulses, vertical sync pulses and equalizing pulses only.

Composite Video: A single video signal that contains brightness, color, and timing information. Composite video is a video stream that combines all of the pieces required for displaying an image into one signal, thus requiring only one connection. NTSC and PAL are examples of composite video. Both are made up of active video, horizontal sync, horizontal blanking, vertical sync, vertical blanking, and colorburst. RGB is not an example of composite video, even though each red, green, and blue signal may contain sync and blanking information, because all three signals are required to display the picture with the right colors.

Composition: The controlled ordering of elements in a scene designed to provide the strongest artistic arrangement and the most effective communication of a central idea. See also Dynamic Composition and Static Composition.

Compound Asset: A media product comprised of more than one element. Also a shot from a film that contains video, audio, musical and sometimes computer-generated materials and elements.

Compress: Reducing a file to a smaller size without altering the substantive information by eliminating the redundant data.

Compressed Time Division Multiplex (CTDM): A method of processing chrominance signals for recording. When component video signals are recorded, both of the two chrominance signals (R-Y, B-Y) are time-compressed to half, multiplexed, and recorded on a single track one after the other.

Compression Algorithm: Mathematical formulas used in compression to set the rules that make a file smaller and easier to deliver over the Internet.

Compression Artifacts: Compacting of a digital signal, particularly when a high compression ratio is used, may result in small errors when the signal is decompressed. These unwanted defects are known as artifacts. Artifacts may resemble noise or edge busyness or may cause parts of the picture particularly fast moving portions to be displayed with the movement distorted or missing.

Compression Ratio: Comparison of file sizes between the uncompressed and compressed files.

Compression: The process of removing redundancies in digital data to reduce the amount that must be stored or transmitted. Lossless compression removes only enough redundancy so that the original data can be recreated exactly as it was. Lossy compression sacrifices additional data to achieve greater compression.

Compulsory License: The right of cable systems and certain other delivery media (i.e., Microwave) to use copyrighted material for a governmentally set fee, without negotiating a price.

Computer Graphics: The creation of still and animated video through the use of computer-type equipment.

Computer Platform: A basic desktop computer. Many editing and special effects hardware and software products use a computer as the basis of their systems.

Computer-generated Imagery (CGI): Images which are created digitally, usually through computer modeling with wire frame objects.

Condensation: Moisture usually on the head drum, which cases the tape to stick to the drum, resulting in damaging the tape and possible malfunction of the recorder.

Condenser Microphone: The simplest type of microphone in which the capacitance (electrical charge) is varied by sound, causing movement in one plate (diaphragm) in relation to a fixed backplate.

Conductor: A person who directs the orchestra's performance of the score, often the composer.

Confidence Heads: VCR audio and video playback heads that immediately play back signals after they are recorded. In this way signals can be monitored for quality as recordings are being made.

Conflict: A basic psychological ingredient in news and dramatic productions which has the power to capture and hold audience attention.

Conform: Making the final edit according to a prepared scheme such as a rough cut or EDL. EDLs can be used to directly control conforming in an on-line edit suite (auto-conforming). The time to conform varies widely, from a tape suite which takes much longer than the finished program running time, to a true random access suite that takes only a moment; the latter still allowing any adjustments to be easily made. Progressive versions of a film in the editing stage are known as conformations, often identified by date. Conformations are only of any significance on a large production where different editing departments should be sure to be working with the latest conformation.

Console: The display television station on the AS/400 that is designated during system configuration to control certain system functions, like messages or backups, and to monitor system operations.

Consortium: A partnership of a number of TV television stations or productions facilities that exchange programming or services. They are generally linked by satellites.

Constant Angular Velocity (CAV): Refers to analog video laser disks which rotate at a constant speed (1800 RPM) to play the program, where about 30 minutes of program fit on a side, and where the ability to do slow motion or single frames is easily accomplished and can be done by any player. One revolution corresponds to one video frame. If you examine the disk surface, you can see a bow tie texture pattern that represents the two interlaced fields and the vertical retrace intervals. There are some cheap tape recorders (Mini-cassette and 3 inch open reel styles) which use the CAV idea, using the take up spool only to pull the tape. As the recording or playback progresses, the tape speed past the heads increases due to the build-up of tape on the take up spool.

Constant Bit Rate: See CBR.

Constant Linear Velocity (CLV): Refers to analog laser disks where the rotational speed varies (1800 to 600 RPM) so that the amount of track circumference spanned by the laser beam is the same for each video frame. Since the circumference of each revolution of the spiral track near the outer edge (later in the program) is greater, the disk is slowed down as the program progresses. A CLV disk holds about an hour's worth of program material on a side.

Construction Coordinator: Also called a Construction Manager, the duties of this position are financial responsibilities including budgeting, tracking costs, generating reports, etc. Through drawings, a construction coordinator is directed artistically by the Production Designer and Art Director to produce their "vision" in three dimensions. Also responsible for the physical integrity of the structures built by the construction department.

Consumer Electronics Association (CEA): An industry group that represents manufacturers of consumer electronics products.

Consumer Electronics Manufacturers Association (CEMA): A subsidiary of the Electronics Industry Association (EIA).

Consumer Equipment: Relatively inexpensive audio and video equipment intended for "home" or non-professional use.

Contact Printing: The method used by the lab to copy film. A contact print is made on a machine called (sensibly enough) a Contact Printer, in which the original film and unexposed print stock are sandwiched together, emulsion against emulsion, and are run at a constant speed past a light which shines through the original, exposing the print stock with the same image. All workprints, answer prints and release prints are contact prints. The only other type of printing is Optical Printing, which is usually done to add an effect or to blow up or make a reduction print.

Contact: A person who knows about an event and its location and can assist the production team during a telecast.

Content Analysis: An empirical method of analysis which selects a specific textual component, counts and codes the number of occurrences of this component into a statistical form, resulting in quantifiable data which usually cannot be interpreted beyond the data itself.

Content Management Information (CMI): General information about copy protection and allowed use of protected content. Includes CCI.

Content Scrambling System (CSS): In DVD-Video, an

encryption scheme designed to protect copyrighted material that resides on a disc by periodically scrambling the data using encryption keys.

Contextual Criticism: A form of criticism that evaluates a film within the context in which it was created and/or screened.

Contingency: A designated amount of a budget which is added in anticipation of potential cost overruns.

Contingent Compensation: Form of compensation received by a writer, after the writing services have been completed, if he is awarded writing credit for the project. The contingent compensation may include a production bonus, net profits, reserved rights, and/or additional payments in the event of a film or television sequel, remake, or spinoff.

Continuity Editing: Editing that creates action that flows smoothly across shots and scenes without jarring visual inconsistencies. Establishes a sense of story for the viewer.

Continuity Person: The person in a production responsible for maintaining consistency in all details from one shot to the next, including action, lighting, props, and costumes.

Continuity Report: A detailed list of the events that occurred during the filming of a scene. Typically recorded are production and crew identification, camera settings, environmental conditions, the status of each take, and exact details of the action that occurs. By recording all possible sources of variation, the report helps cut down continuity error between shots or even during reshooting.

Continuity: The seamlessness of detail from one shot to another within a scene. Continuity refers particularly to the physical elements, rather than to the choices in coverage that can result in a lack of seamlessness. Elements of continuity include any actions of the actor, the placement of props, the lighting, the costumes, and so on.

Continuous Printing: A method of printing in which both

the original and the raw stock move continuously past the point of exposure.

Continuous White Balance: Automatic camera white balance mode that makes moment-by-moment adjustments in response to changing light conditions.

Contouring: One of the initial steps in the application of make-up designed to emphasize or deemphasize certain facial features and dimensions. Contouring is also an image artifact caused by not having enough bits to represent the image. The reason the effect is called "contouring" is because the image develops vertical bands of brightness.

Contract: An order by the client for the broadcast of Spots. It contains information about desired timing and context of the spots as well as the billing information. The contract usually serves as the legal agreement between an agency or advertiser and the television station. After an order has been entered into the system, it is printed on a confirmation or contract, which serves as a printed verification of the contractual data stored in the computer.

Contrast Control: A control used to change the amplitude of the video signal in viewfinders receivers, and monitors, but not affecting the output of cameras or other video sources.

Contrast: The difference in illumination between the brightest and darkest parts of a scene or picture.

Contribution Quality: Describes a level of encoding or compression for use in a production environment. Contribution quality has the following attributes: short latency, multiple generations with minimal degradation, transparent concatenation with distribution level compression encoding. MPEC 2@ 4: 2: 2 is an example of contribution quality compression.

Control Area: A part of the lead-in area on a DVD containing one ECC block (16 sectors) repeated 192 times. The repeated ECC block holds information about the disc.

Control Room: A room associated with a production

studio in which a director, technical director, audio engineer and others work to coordinate and control various sources of audio and video during a production.

Control Track Editing: A system of editing in which the machine counts the control track pulses on a videotape to establish edit points and pre-roll times. Accuracy problems result with very slow or very fast shuttle speeds.

Control Track Editing: Equipment that uses an electronic count of control tack pulses rather than SMPTE/EBU time code numbers for its cuing and editing functions.

Control Track: A recorded track used to adjust or manipulate a recording or playback device. Generally, these are used to maintain consistent playback speed of video and/or sound recorders consistent with the recorded speed.

Control-L: Sony's wired edit control protocol, also called LANC (Local Application Control), which allows two-way communication between a camcorder or VCR and an edit controller such as the Thumbs Up. Control-L allows the controller to control the deck (fast forward, play, etc.) and also allows the controller to read the tape position (tape counter) information from the deck.

Control-M: Panasonic's wired edit control protocol. Similar to Control-L in function but not compatible. Also called Panasonic 5-pin edit control.

Control-S: Sony wired transport control protocol which duplicates a VCR's infra-red remote transport control (play, stop, pause, fast forward and rewind). Unlike Control-L, Control-S does not allow the controller to read tape counter information.

Control-T: Similar to Control-L but allows multiple units to be controlled. Not used in current equipment.

Convergence: The accuracy of the positions of the red, green, and blue beams of a color monitor or projector. Also the adjustment of the red, green, and blue electron beams in a monitor or video projector to align the red, green and blue images.

Conversion Filter: A lens attachment designed to convert the color temperature of outdoor light to indoor color temperature or vise-versa.

Converter: A device that translates cable signals into television signals and allows the viewer to select individual channels.

Convertible Camera: A video camera that can be quickly switched between use in the studio and in the field.

Coocoloris: A fancier way of saying Gobo or Cookie.

Cookie: A flat board, like a flag, but full of irregular holes used for creating a pattern of shadows when put in front of a light.

Cool: A television picture that has a slightly blue or green cast.

Cooperatives: Groups of broadcast television stations and production facilities that exchange programming typically news segments.

Co-producer: A producer who has equal responsibility for the completion of a project.

Copy Protection Device: A system designed to foil attempts to make illegal copies of copyrighted audio or video recordings.

Copy Stand: Device for holding flat artwork and a camera. Used for photographing artwork under controlled conditions.

Copy: Literally, material to be read or produced.

Copyright: The exclusive legal rights to perform or sell a song, book, script, photograph, etc. To use copyrighted material (e.g., a piece of music) in a TV program, a fee or royalty must be paid the copyright holder. If there is no copyright the material may be used for free and is said to be in the public domain.

Core: A plastic cylinder on which film is wound for transport or storage.

Coring Out: The process of removing video noise by converting the analog signal into a digital signal and then back again.

Corner Insert: A square or rectangular key insert into a segment of a background picture. Generally used in news to show graphics behind a newscaster.

Corner Pinning: Method for defining the position (keyframe) of a picture by "pinning" its four corners: usually in a DVE or graphics system. With the appropriate 3-D mathematics the four corner pins can define the picture's position and orientation in 3-D space. The method is a highly efficient alternative to the common, and often tedious, positional and rotational DVE controls, making light work of placing pictures or objects accurately into scenes. With moving video a series of corner-pinned key frames can set up a track of an object.

Corrected Print: Same as a Timed Print.

Corrections: Further changes in the timing of a print are known as corrections.

Corrective Make-up: Make-up intended to significantly change appearance generally to hide certain original facial features.

Cost Per Mill (CPM): The advertising rate charged to TV sponsors, which is quantified per thousand viewers. "Mil" equals "thousand," from the Latin word mile. Thus, the CPM is the cost per thousand viewers.

Cost Per Minute: Production costs divided by total production time.

Cost Per Point (CPP): The media cost of reaching one percent (one rating point) of a household or demographic.

Cost Per Viewer (CPV): Production costs divided by the number of viewers.

Cost Per Measured Results: Comparing the costs of a production with the intended results. In commercial television this could consist of comparing increased sales revenue with the costs of doing a commercial.

Costing Out: The process of figuring out the costs of a production.

Costume Designer: Individual who designs the costumes for a film.

Costume Supervisor: Individual responsible for handling the costumes worn by actors.

Count Down: Visual cues to talent, usually from ten seconds, to start or to the end of a production or segment.

Countdown Leader: Similar to Academy Leader in film, a visual, (and sometime aural) second-by-second countdown from 10 to 3 seconds. Used cue the beginning of segments for playback.

Counter Programming: Competitive scheduling, placing a program which appeals to a certain type demographic against other programs which appeal to a different demographic during the same time period.

Counter: A device for measuring lengths of film by counting the number of frames.

County Size: Market Sections defined by Nielsen and divided into four categories based upon the 1990 census. "A" County**:** The counties belonging to the 21 largest metropolitan areas. "B" County**:** All counties, not counted under "A," that are in the metropolitan areas with over 85,000 households. "C" County**:** All counties, not included under "A" or "B," that have over 20,000 households or are in metropolitan areas with over 20,000 households. "D" County**:** All remaining counties.

Coupler: A chemical incorporated into the emulsion of color film stocks which produces a dye image associated with the developed silver image.

Cover Set: A location which is kept in reserve to serve as an alternate shooting site in case the chosen shooting site is unusable. It is most commonly used in the context of shooting planned for an out of doors location.

Cover Shot: An establishing wide-angle or long shot of a set used both to establish the relationships between subject matter in a scene and to momentarily cover problems with lip sync or mismatched action.

Coverage Area: A geographic area within which a television station or radio television station can be received. The FCC requires that each broadcast licensee produce a coverage map.

Coverage: An indeterminate number of more detailed shots which are intended to be intercut with a master shot or scene. During the process of photographing a scene for a media program, coverage is a term given to the acquisition of photographed scenes other than the master scene. This would include close-ups, over the shoulder, extreme wide shots or reaction shots. Coverage is often not photographed at the same length as the original master scene being covered.

CP Filters: Color printing filters made in precise density values for the colors; yellow, cyan, and magenta.

CPM: See Cost Per Mill.

CPP: See Cost Per Point.

CPPM (Content Protection for Prerecorded Media): Copy protection for DVD-Audio.

CPRM (Content Protection for Recordable Media): Copy protection for writable DVD formats.

CPS_TY (Copyright Protection System Type): In DVD-Video, an 8-bit (1 byte) value stored in the CPR_MAI that defines the type of copyright protection system implemented on a disc.

CPSA (Content Protection System Architecture): An overall copy protection design for DVD.

CPTWG (Copy Protection Technical Working Group): The industry body responsible for developing or approving DVD copy protection systems.

CPV: See Cost Per Viewer.

CR: See Camera Roll.

Crab Dolly: A camera mount with steering controls that can move the camera in any direction.

Crabbing: Moving a camera sideways or in arc on a crab dolly. Also referred to as Trucking.

Cradle Head: A heavy, cradle-shaped camera pan head consisting of two half-circle elements.

Crane Shot: A shot taken from a studio crane, a large mechanical arm that can move the camera and its operator smoothly and noiselessly in any direction.

Crash Edit: An edit that is technically wrong. On tape this may be caused by a number of factors including tape damage and discontinuities in control track, field sequence or time code across an edit. These may result in a picture disturbance on replay. Such technical problems need not apply to disk-based operations.

Crawl: The movement of credits or other graphic material across the screen.

Creative Compromise: Including and excluding elements in a production based primarily on budget concerns. An example would be eliminating a top star in favor of enhanced special effects or an exotic production location.

Creator: The writer or the primary creative force behind a movie, series, or group of characters.

Credit Arbitration: Process run by the Writers Guild of America in which disputes concerning the award of credit (as in "Story by, "Screenplay by," "Teleplay by," and "Written by") are decided. The method in which these decisions take place has the WGA sending all drafts of the disputed work to three separate individuals; separately and without knowledge of each other, they decide which writer deserves the award of credit. When two of the three individuals agree on the award of credit, the decision is considered final.

Credits: Listings of all those involved in making a program, usually appearing at the end of a television program or film.

Creepy Crawlies: This term refers to a specific image artifact that is a result of the NTSC system. When the nightly news is on, and a little box containing a picture appears over the anchorperson's shoulder, or when some computer-generated text shows up on top of the video clip being shown, get up close to the TV and check it out. Along the edges of the box, or along the edges of the text, you'll notice some jaggies "rolling" up (or down) the picture. That's the creepy-crawlies. Some people refer to this as zipper because it looks like one.

Crewmember: A collective term for anyone involved with the production of a movie who does not appear in the movie. This term is usually used to refer to the more subordinate members of a production team.

CRF: See Cutting Room Floor.

CRI: See Color Reversal Intermediate.

Cribbing: Short pieces of lumber which are used for various grip purposes.

Critic: A person who publishes a review of a movie from either an artistic or entertainment point of view.

Critical End!: What to label the film can when turning it in at the lab when the roll ran out during a very important shot and you want to make sure you get every last frame possible.

Crop: To trim and remove a section of the video picture in order to make it conform to a different shape. Cropping is used in the pan & scan process, but not in the letterbox process.

Cross Back Lights: Using two back lights a few feet apart in such a way that when viewed from the camera position the light very slightly wraps around the sides of the subject.

Cross Color: Rainbow swirls amongst or swamping out pinstripes and other fine detail, caused by small amounts of luminance signal left behind in the signal going into the color circuits. This situation is the result of imperfect Y/C separation by the comb filter (or notch and bandpass filters in less expensive TV sets).

Cross Cutting: Cutting back and forth quickly between two or more lines of action, indicating they are happening simultaneously. See also Parallel Action.

Cross Fade: The process of bringing down one audio or video source while simultaneously bringing up another.

Cross Luminance: Crawling or hanging dots where color patches meet, or silk screen effects, caused by small amounts of color signal going into the luminance circuits. This situation is the result of imperfect Y/C separation by the comb filter (or notch and bandpass filters in less expensive TV sets). This occurs when the video decoder incorrectly interprets chroma information (color) to be high-frequency luma information (brightness).

Cross Modulation Test: Sometimes called "cross mod" for short. This is a test the Mixing House will do in conjunction with the lab you plan to use to make sure the optical track is exposed and developed for optimal sound quality.

Cross Modulation: A condition when one signal erroneously modulates another signal.

Cross Platform: The ability for software to work on two or more computer systems, such as PC and Macintosh.

Cross Processing: A technique used much more by still photographers. Cross processing is the use of color reversal film stock to be developed as a negative. A positive print struck from that negative will have strange and rich colors, intense contrast and on overall yellowish hue.

Crossing the Line: A 180-degree or more shift in camera position resulting in a reversal of on-screen action or audience perspective. "The line" is also referred to as the Imaginary Line and the Action Axis.

Crossover: The frequency at which a signal is split in order to feed separate parts of a loudspeaker system.

Crosstalk: The interference between two audio or two video signals caused by unwanted stray signals. In video, crosstalk between input channels can be classified into two basic categories: luminance/sync crosstalk; and color (chroma) crosstalk. When video crosstalk is too high, ghost images from one source appear over the other. In audio, signal leakage, typically between left and right channels or between different inputs, can be caused by poor grounding connections or improperly shielded cables.

CRT: See Cathode Ray Tube.

CRTC (Canadian Radio/Television and Telecommunications Commission): The CRTC is the regulatory agency governing the broadcast industry in Canada. It is roughly the Canadian equivalent of the FCC.

Crystal Sync: Specifically, a way of recording Sync Sound where the camera runs at correct speed with a quartz crystal-governed motor, and tape recorder records its pilottone using a built-in quartz crystal pilottone generator. The crystal is much like the kind used in a quartz watch. Unlike cable sync, the camera and tape recorder are not attached.

CSA: See Casting Society of America.

CSS: See Content Scrambling System.

C-stand: A type of light stand with fixed legs that swing out, or together when not in use, usually equipped with an arm, and typically used to hold a flag.

C-Stand: See Century Stand.

CTB (Color Temperature Blue): Color correction gels used in lighting to convert the color temperature from tungsten to daylight. They come in gradients: Quarter Blue, Half Blue, and Full Blue.

CTDM: See Compressed Time Division Multiplex.

CTL: Control signal in the form of regular pulses recorded along a longitudinal track on the videotape. By counting these pulses, it is possible to determine the number of frames, and hence the tape's running time. Used mainly to adjust the tracking position of video heads, and to achieve time code continuity in continuous recording.

CTO (Color Temperature Orange): Color correction gels used in lighting to convert the color temperature from daylight to tungsten. They come in gradients: Quarter Orange, Half Orange, and Full Orange.

CU: See Close-up.

Cube Flip: A special effect in which pictures appear to be glued to the sides of a cube.

Cue Cards: Large white cards used by on-camera talent to help them remember dialogue. Black markers are normally used to write material in large letters.

Cue Channel: See Audition Channel.

Cue Patch: A piece of self adhesive magnetic or metallic material placed on the edge of film as a means of actuating a printer light change or an automatic stop on a projector.

Cue Sheets: A road map, of sorts, for the mixer to find the sounds on the tracks during the mix. It is laid out as a grid with each track forming a column and time moving ahead in rows measured in 35mm footage (even if the film is 16mm you must convert the footage to 35mm).

Cue Track: A videotape audio track that can be used in several ways: to record SMPTE/EBU time code, to record the PL line conversation or in-house identification or cues, or for a second audio track.

Cue: Signal to start or stop an operation or to move to a predetermined point in program material. On tape this will involve spooling and, maybe, a reel change. On disk it can be instant.

Cueing: A term with a broad range of meanings depending on the particular context. For Voice-Over Narration or Dialogue Replacement: The marking of the cue point in a way which will permit a signal to be given to the 'talent' to begin each element of work at the appropriate time. In General: Any system used by a second person to signal 'talent' that recording should begin.

Cultural Studies (Ethnography): A critical approach which argues that viewers decode television texts based on their specific ideological position in society; it looks at the interaction between the ideological discourses of the text and those of the viewer.

Cup Blocks: Wooden blocks with a dish or indentation in the center which are used to keep the wheels of light stands from moving.

Curvi-linear Effects: DVE picture manipulations that distort the flat TV image in a curved fashion. This may bend the picture, hence page turns, or roll it into any number of shapes.

Cut: In filming, to change from one shot to another immediately. In directing, called by the director to stop action by the performers, camera, and audio equipment. In film editing, to eliminate unwanted portions, visual or audio, of a film.

Cutaway: The use of a shot that is not part of the primary action but which is relevant to it and occurring at the same time. In an interview a cutaway is commonly used to show the interviewer's reaction to what is being said.

Cut-in Shot: The use of a close-up shot of something within a scene in order to show added detail.

Cut-in: A television station substitution of a local spot that will "cover up" a spot being fed to the television station from another source, such as the network (also known as a Rollover). It may be regional messages in a network show or a local news bulletin in a national news show.

Cut-off: The latest time a spot is allowed to run. Cut-off time is determined by the client and is generally applied across the board. Spots running after cut-off usually will not be paid for.

Cutout: See Matte.

Cutting in the Camera: Shooting scenes in the desired sequence so that little or no editing is required. Often used in covering news stories where there will not be enough time for editing.

Cutting on Action: Editing two shots at a point where a gesture or movement in the first one is not yet completed and where a gesture or movement in the second one has already begun.

Cutting on the Beat: Editing shots in time with music. Generally cuts begin at the start of a measure of music or on any beat within a measure.

Cutting Room Floor (CRF): Moving material to CRF in Quantel editing systems removes it from the work area, but not from the disk store.

CVBS: Composite Video Baseband Signal.

Cyan: A blue-green color which is the complementary color to red.

Cyberpunk: A subgenre of science fiction that typically has elements which include a futuristic tone, massive urban areas in decay and poverty, partial environmental collapse, extremely powerful business corporations, random street gang violence with the overall presence of extremely powerful computer, robotic and information technology. Blade Runner is considered the definitive cyberpunk movie.

Cyclorama Lights: Row lights for evenly illuminating a cyclorama or other background.

Cyclorama: A large, curved, seamless, background generally white or light gray.

D

D/A: Digital to Analog, as in the audio D/A converter in the Plum breakout box.

D1: D2: D3: Digital video recording and playback formats. The D1 system uses component video while the D2 and D3 systems use composite video. By using fully digitized video in recording and playback, many problems such as generation loss and distortion are minimized or eliminated. The digital formats use mainly a 19mm wide magnetic tape (3/4").

D9 Video: Digital videotape format using CCIR 601, 4: 2: 2 video compressed about 3: 1. The same as "Digital S."

Dadaism: An avant-garde movement in the arts during the 1920s which, like surrealism, stressed the unconscious and the irrational in human experience, incongruity, spontaneity, and irreverent wit.

DAE: See Digital Audio Extraction.

Dailies: The first positive prints made by the laboratory from the negative photographed on the previous day. It also now refers to video which is transferred from that original negative. Also known as Rushes.

DAM (Digital Asset Management): DAM is the cataloging and storing material on a digital system in such a way that authorized people can search, locate and retrieve, manipulate and return all or part of the contents. In an entertainment company, the system must accommodate text, drawings, 2-D and 3-D models, photographs and high-quality video and audio.

Dance Floor: A floor built of 3/4 inch plywood which is usually covered with masonite to provide a smooth surface for free-form dollying.

DAT (Digital Audio Tape): Audio recording and playback format developed by Sony, with a signal quality capability surpassing that of the CD.

Data Area: The physical area of a DVD disc between the lead in and the lead out (or middle area) which contains the stored data content of the disc.

Data Compression: A technique that provides for the transmission or storage, without noticeable information loss, of fewer data bits than were originally used when the data was created.

Data Management: Organizational system for including descriptive information in a media file, such as author, copyright, bit rate, etc.

Data Rate: The amount of data moved over a period of time, such as 10MB per second. Often used to describe a hard drive's ability to retrieve and deliver information.

Data Search Information (DSI): Information for Fast Forward/Fast Backward and seamless playback. This is real time control data spread throughout the DVD-Video data stream. Along with PGCI, these packets are part of the 1.00 mbit/sec overhead in video applications (Book B). These packets contain navigation information which makes it possible to search and maintain seamless playback of the Video Object Unit (VOBU). The most important field in this packet is the sector address where the first reference frame of the video object begins. Advanced angle change and presentation timing are included to assist seamless playback.

Datacasting: While broadcasting in HDTV or multicasting in SDTV, digital technology allows broadcasters to use leftover bandwidth to transmit additional program material or non-program related resources, such as video, audio, text, graphics, maps and services, to specially equipped computers, cache boxes, set-top boxes, or DTV receivers. This is called datacasting. DTV's broader bandwidth channel allows information to be downloaded at a transmission rate currently 600 times that of a personal computer modem.

DAW (Digital Audio Work Station): A computer-based recording and editing machine used for manipulating sounds.

Day Out of Days: A form designating the workdays for various cast or crewmembers of a given production.

Day-for-night: This means shooting night scenes during the day. To this properly you have to use camera filters and underexposure. If you are shooting on B & W film use a red or yellow filter to darken the sky. With color film use an ND or Neutral Density filter. Shooting Day-For-Night works best on really sunny days, either early in the morning or late in the afternoon, because the shadows are hardest then.

Daylight Balanced: The color temperature of daylight which is 5,400K on the color temperature scale (it does vary during the day, being higher at noon and lower in the earlier and later parts of the day). Color film for outdoor shooting is balanced for daylight, otherwise the image would appear blue in hue. If daylight balanced film is used indoors without a correction filter the image will have a orange hue.

Daylight Spool: An aluminum spool holding 100 feet of film with solid, opaque sides, painted black, which will protect the film from becoming completely exposed when loading a camera in daylight. The name daylight spool comes from the fact that the film may be loaded without total darkness. There are also 400 foot daylight spools, but these are very rarely used as they do not always work very well in a magazine.

Daypart: The division of a broadcast day (television or radio) into individual parts, in network television an example might be: 1) Morning M-F 7A-9A 2) Daytime M-F 10A-4P 3) Early Fringe M-F 4P-7P 4) Prime Time Access M-Sat 7P-8P 5) Prime Time M-Sat 8P-11P & Sun 7P-11P 6) Late News M-F 11P-11: 30P 7) Late Fringe M-F 11: 30P-1A.

De-emphasis: Also referred to as post-emphasis and post-equalization. De-emphasis performs a frequency-response characteristic that is complementary to that introduced by pre-emphasis.

Deep Focus Photography: A cinematographic technique which keeps objects in a shot clearly focused from close-up range to infinity. Also called Pan Focus Photography.

Deep Focus Shot: A shot in which both the foreground and the background are in focus. In other words, a shot with exceptional depth of field.

Deep Space Blocking: A type of blocking associated with single-camera productions, particularly those shot on location. The depth of the "set" is emphasized by the ability of one actor to be positioned near the camera and another far away; the actors may move toward one another, or participate in independent actions.

Defamation Per Quod: As opposed to defamation per se, statements which are deemed slanderous or libelous only when supplemental (generally, widely-known information) is considered along with the statement in question.

Defamation: A legal term relating to the utterance of slanderous words, especially false and malicious words, that injure a person's reputation.

Deferred Compensation: Money to be paid from funds that, for a writer, are generally the net profits. Rarely will the writer ever see such compensation.

Definition: The aggregate of fine details of a video image on-screen. The higher the definition, the more discernable the details.

Defocus: Optical or digital effect where the whole frame of video or some part of it appears to be significantly blurred. For instance, a single white pixel will spread gradually over several of its neighbors. Similar to, but, not the same as anti-aliasing.

Degausser: A device used to erase recordings on magnetic tapes and films or to demagnetize magnetic recording heads.

Deinterlace: The process used to convert interlaced frames to progressive scan frames. This process properly prepares native television signals for playback on progressive scan devices, such as computer screens.

Delay Correction: When an electronic signal travels through electronic circuitry or even through long coaxial cable runs, delay problems may occur. This is manifested as a displaced image and special electronic circuitry is needed to correct it.

Delayed Broadcast (DB): The replaying of an original broadcast by a television station at a time later than generally scheduled.

Delegation Control: Switcher control which can assign different functions to the buses.

Delta Picture: A video picture based on the changes from the picture before (or after) it. MPEG P pictures and B pictures are examples. Also called Delta Frame.

Demodulation: The process of recovering an original signal from a modulated carrier.

Demographics: The characteristics of an audience, usually broken down in terms of age, gender, income, race, etc.; used with ratings to set advertising rates.

Demultiplex: The separation of two or more signals that were previously combined by a compatible multiplexer and transmitted over a single channel.

Denoument: The concluding scenes of a movie where the story elements are finished and the characters' status after the climax is shown.

Density: A factor which indicates the light-stopping power of a photographic image.

Depth Fog: 3-D effect where objects fade into a background color, or fog, as they go into the distance. Fog color and distances can be set. The technique is used primarily to reduce visible polygons, because polygons completely beyond the fog are clipped (not drawn).

Depth of Focus: The range of distances from the rear of the lens to the face of the pickup tube in which acceptable focus can be achieved.

Depth Staging: Arrangement of objects on a set so that foreground, middle ground, and background are each clearly defined on a TV screen.

Descrambler: An electronic circuit that decodes encoded TV signals.

Descriptive Video Services: See DVS.

Designated Market Area (DMA): Groups of counties that are assigned to local television markets based upon television stations in those markets which account for the plurality of counties total viewing hours. Updated annually by Nielsen Television station Index, there are 211 DMA's in the country, none of which overlap.

Designer: Someone who creates plans for visual aspects of a production.

Desktop Video: Video editing and production using standard desktop computing platforms running with add-on video hardware and software.

Deterministic: A process or model whose outcome does not depend upon chance, and where a given input will always produce the same output. Audio and video decoding processes are mostly deterministic.

Deuce: A 2K fresnel lighting unit.

Developing: The chemical process which converts a photographic exposure into a visible image.

Development: In the entertainment industry, development usually refers to the act of creating and refining a story idea to make it ready to produce. The development process often additionally refers to the casting of major talent for the production including the Director, Talent and even some "below the line" talent such as key crew members.

Device Interface: A conversion device that separates the RGB and sync signals to display computer graphics on a video monitor.

Dew Light: A warning light that protects a VCR by indicating moisture or condensation present inside the unit.

DFU: See Data File Utility.

DGA: See Directors' Guild of America.

Diagnostics: Tests to check the correct operation of hardware and software.

Dialogue Editor: A sound editor who specializes in editing dialogue.

Dialogue Track: A soundtrack which carries lip sync speech.

Dialogue: Speech among characters, which does not usually address the viewer. Also, a type of interview in which the voices of the interviewer and the interviewee are both heard, and both persons may be visible on camera.

Diaphragm: A lens aperture or iris made of overlapping metal leaves that controls the amount of light passing through the lens.

Diary: A questionnaire in which the respondent is asked to furnish a written record of their television viewing activity over a specified period of time (usually one week). This methodology is prevalent in local (spot) television ratings.

Diascope: A type of internal test pattern built into a zoom lens which can be flipped into position to register and align pickup tubes.

Diachronic Filter: An optical device which passes or reflects only certain wavelengths of light. Normally used in three-CCD color cameras to separate red, green, and blue.

Diegetic Sound: Dialogue, music and sound effects that

occur in the diegetic space of the television program. I.e., sound that is part of the characters' world.

Diegetic Space: The physical world in which the narrative action of the television program takes place.

Difference Keying: Method used to derive a key by mathematically subtracting the values of one picture from another. Typically this is applied using a locked-off camera. The first shot may be an empty scene and the second includes action (e.g. a bull charging down a street). The difference between the two shows as the action (the bull) and a key can be made from that.

Differential Gain: A term that describes how much the color saturation changes when the luma level changes. For a video system, the better the differential gain—that is, the smaller the number specified—the better the system is at figuring out the correct color.

Differential Phase: A term that describes how much the hue changes when the luma level changes. For a video system, the better the differential phase—that is, the smaller the number specified—the better the system is at figuring out the correct color.

Differential Rewind: A device designed to permit simultaneous winding of film on more than one reel at a time even though the diameters of the rolls are unequal.

Diffused Light: Soft light. Light which casts an indistinct, soft shadow.

Diffusion Filter: Filter that decreases image sharpness. Used for smoothing out skin blemishes and for dreamlike effects. Filter consists of polished glass with one side smooth and the other slightly rippled which causes light passing through it to be deflected slightly to decrease image sharpness.

Digital 8: Digital 8 compresses video using standard DV compression, but records it in a manner that allows it to use standard Hi-8 tape. The result is a DV "box" that can also play standard Hi-8 and 8 mm tapes. On playback, analog tapes are converted to a 25 Mbps compressed signal available via the i-Link digital output interface. Playback from analog tapes has limited video quality. New recordings are digital and identical in performance to DV; audio specs and other data also are the same.

Digital Asset Management: See DAM.

Digital Asset: Any digitally encoded media product that has potential value in any format.

Digital Audio Extraction (DAE): Reading digital audio data directly from a CD audio disc.

Digital Betacam: Introduced by Sony in 1993, a version of Betacam that offers the advantages of digital quality.

Digital Broadcast Satellite (DBS): The general term for 18-inch digital satellite systems.

Digital Component Video: Digital video using three separate color components, such as Y'CbCr or R'G'B'.

Digital Composite Video: Digital video that is essentially the digitized waveform of NTSC or PAL video signals, with specific digital values assigned to the sync, blank, and white levels.

Digital Compositing: A technique whereby separately filmed components are combined through digital editing.

Digital Compression: A process of translating video images into a digital code which takes up less transmission space than the original signal would have. This allows more channels per satellite transponder: from four-to-one for live video to eight-to-one for film.

Digital Disk Recorder (DDR): A digital video recording device based on high speed computer disk drives. Commonly used as a means to get video into and out from computers.

Digital Editing: Editing a portion of a movie by digitizing one or more frames and altering them.

Digital Fingerprint: A security record that tells who

entered the asset management system, how long they were there, and what files they accessed.

Digital Linear Tape (DLT): A digital archive standard using half-inch tapes, commonly used for submitting a premastered DVD disc image to a replication service.

Digital Media Infrastructure: Infrastructure technology and processes that enable the encoding, management and distribution of audio and video on the Internet.

Digital Negative: A term used to describe a string of digital words, bytes and files that describe the necessary components to reproduce a media image. The digital negative then becomes the master storage medium for the program material including all components such as sound and image.

Digital News Gathering (DNG): Electronic news gathering (ENG) using digital equipment.

Digital Recording: A method of recording in which samples of the original analog signal are encoded on tape or disk as binary information for storage or processing. The signal can then be copied repeatedly with no degradation.

Digital Signal Processing: Use of mathematical expressions, such as algorithms, to modify digital signals.

Digital Signal Processor (DSP): A specialized processor often used to accelerate the rendering of video effects. The advent of high speed processors such as 166 MHz Pentiums has largely eliminated the need for DSP based video rendering.

Digital Signal Processor: An integrated circuit that performs functions such as reverb, filtering and equalization on a digitized waveform.

Digital Sound: A technology (e.g., CDs) that converts sound into numbers; this allows computers to process and/or change the recorded sound. It has been replacing analog sound processes (e.g., vinyl albums and audio cassettes).

Digital Television: The umbrella term for the new broadcasting system adopted by the Federal Communications Commission in 1996. Analog television, which was developed in the 1940s, receives one continuous electronic signal. In contrast, DTV works on the same principle as a computer or a digitally recorded compact disc. It uses binary code, a series of ones and zeros, rather than a continuous signal. Permits HDTV, multicasting, and enhanced TV.

Digital Theatre Systems: A company which has produced a digital soundtrack standard. Competitors include Dolby Digital and SDDS.

Digital Transmission Content Protection: An encryption method (also known as "5C") developed by Sony, Hitachi, Intel, Matsushita, and Toshiba for IEEE 1394 interfaces.

Digital VCR: Digital VCRs are similar to analog VCRs in that tape is still used for storage. Instead of recording an analog audio/video signal, digital VCRs record digital signals, usually using compressed audio/video.

Digital Vertical Interval Time Code (DIVTC): This system digitizes the analog VITC waveform to generate 8-bit values. This allows the VITC to be used with digital video systems. For 525-line video systems, it is defined by SMPTE 266M. BT.1366 defines how to transfer VITC and LTC as ancillary data in digital component interfaces.

Digital Video (DV): Any video format that relies on digital technology for recording and/or editing. For example, video recorded with a digital camera or edited on a nonlinear editing system.

Digital Video Broadcast (DVB): Broadcast standard for digital radio and television, using MPEG II compression. DVB is being supported by all European manufacturers and broadcasters.

Digital Video Effects (DVE): Special effects created with digital, computer-based technology. Compare with electronic effects. Common DVE effects include inverting the picture, shrinking it, moving it around within the frame of another picture, spinning it, and a great many more.

Digital Video Noise Reduction (DVNR): Digitally removing noise from video by comparing frames in sequence to spot temporal aberrations.

Digital Visual Interface (DVI): The digital video interface standard developed by the Digital Display Working Group (DDWG). A replacement for analog VGA monitor interface.

Digital Watermark: A technique to protect intellectual property. Bits are spread throughout an image that cannot be seen by the naked eye, but can be read by a computer to identify the rights holder.

Digital Zoom: In contrast to an optical zoom, an electronic method of enlarging a picture creating a zoom effect, generally with a resulting loss of sharpness and detail.

Digital: Information in the form of binary data. Computers are digital machines that use a binary system. That is, at their most basic level, computers can distinguish between just two values, 0 and 1 (i.e., off and on). There is no simple way to represent all the values in between, such as 0.25. All data that a computer processes must be digital, encoded as a series of zeros and ones. Digital representations are approximations of analog events. They are useful because they are relatively easy to store and manipulate electronically.

Digital-S: JVC's digital video compression and recording system which uses SVHS tape.

Digitize: To convert analog video, audio, or both to digital form.

Digitizing Time: Time taken to record footage into a disk-based editing system. The name suggests the material is being played from an analogue source which, with the rapidly increasing use of DVTRs, is not always the case. A better term is "loading."

Dimensional Mergers: A composition problem often caused by lighting or tonal contrast problems where objects at different distances in a scene merge together and cannot be adequately distinguished.

Dimmer: A device for varying power to the lights.

DIN (Deutsche Industrie Norme): An international connector standard. DIN connectors carry both audio and video signals and are common on equipment in Europe.

Dingle: Branches which are placed in front of a light as a cookie would to cut the light and provide a shadow pattern.

Diopter: The part of the viewfinding system of a camera that can be adjusted to compensate for the own particular eyesight, allowing you to see the groundglass clearly.

Dipole: An antenna design with two elements.

Direct Response Advertisers: Advertiser commercials that ask for prompt response from a viewer. Advertisers' commercials show telephone numbers and box numbers necessary to inquire about products and services.

Directing Animator: The animator responsible for creating the key poses or key frames of an animation.

Directional Characteristic: The variation in response or perception for different angles of sound incidence.

Directional Microphone: A microphone that is sensitive along a particular (generally narrow) angle of acceptance. Commonly used to attenuate unwanted off-axis sound.

Director of Photography: A cinematographer who is ultimately responsible for the process of recording a scene in the manner desired by the director. The Director of Photography has a number of possible duties: selection of film stock, cameras, and lenses; designing and selecting lighting, directing the gaffer's placement of lighting; shot composition (in consultation with the director); film developing and film printing.

Director: The principal creative artist on a movie set. A director is usually (but not always) the driving artistic source behind the filming process, and communicates to actors the way that he/she would like a particular scene played. A director's duties might also include casting, script editing, shot selection, shot composition, and editing. Typically, a director has complete artistic control over all aspects of the movie, but it is not uncommon for the director to be bound by agreements with either a producer or a studio. In some large productions, a director will delegate less important scenes to a second unit.

Director's Cut: Contracts under the terms of the Hollywood Director's Guild usually allow 6 weeks for a director to assemble a cut of the movie without studio interference as he or she would like it to be seen. This director's cut is fully edited and has a synchronized soundtrack. This cut is usually not color corrected or density corrected and may not even have the final music and effects tracks. In more recent times the term Director's Cut has taken on a popular meaning that implies a polished final cut of the movie that the director has complete artistic control over.

Directors' Guild of America (DGA): A union for directors and stage managers.

Directory Structure: For the video specification (Book B), this defines a common set of files that must be present on all DVD discs. Components include Root and VIDEO_TS.

Directory: The part of a disc that indicates what files are stored on the disc and where they are located.

DirectShow: A software standard developed by Microsoft for playback of digital video and audio in the Windows operating system. Replaces the older MCI and Video for Windows software.

Disc Key: A value used to encrypt and decrypt (scramble) a title key on DVD-Video discs.

Disc Menu: The main menu of a DVD-Video disc, from which titles are selected. Also called the system menu or title selection menu. Sometimes confusingly called the title menu, which more accurately refers to the menu within a title from which audio, subpicture, chapters, and so forth can be selected.

Discharge-type Lamp: Any one of several types of lights which does not rely on a tungsten-type filament for illumination. Typically, a lamp which creates illumination through the interaction of a high voltage with gas.

Discourse: Socially-based belief structures. The viewer brings discourses to the reading of the television text, which contains discourses that match or clash with the viewer's.

Discrepancy Report: A written account of any technical or operational problems or errors which affect broadcast programming.

Discrete Surround Sound: Audio in which each channel is stored and transmitted separate from and independent of other channels. Multiple independent channels directed to loudspeakers in front of and behind the listener allow precise control of the soundfield in order to generate localized sounds and simulate moving sound sources.

Discrete Time Oscillator (DTO): A discrete time oscillator is a digital version of the voltage-controlled oscillator.

Dish Antenna: A bowl-shaped antenna reflector that concentrates focuses signals toward a receiving element. Typically used to receive satellite signals.

Disk Drive: A sealed computer storage device which utilizes a magnetic disk or, the actual mechanism that turns the disk to read and write digital information to the computer disk.

Display Rate: The number of times per second the image in a video system is refreshed. Progressive scan systems such as film or HDTV change the image once per frame. Interlace scan systems such as standard television change the image twice per frame, with two fields in each frame. Film has a frame rate of 24 fps, but each frame is shown twice by the projector for a display rate of 48 fps. 525/60 (NTSC) television has a rate of 29.97 frames per second (59.94 fields per second). 625/50 (PAL/SECAM) television has a rate of 25 frames per second (50 fields per second).

Display Screen: Similar to a TV screen, the display screen is part of the display television station. It is designed to assist the operator by displaying operator prompts, input data as it is keyed, and output data and operator messages.

Display Size: The actual screen area of a monitor. Generally, the diagonal distance across the screen.

Display Television Station Operator: The person using the system for entering data, file inquiry, and requesting jobs.

Display Television Station: A hardware device which includes a screen to display data and a keyboard to request jobs and enter data. When the system is configured, the display television station can be defined either as a command display television station or as the system console. The display television station designated as the console can also operate as a command display television station.

Display: A visual presentation of data, as seen on a display screen. Display is not synonymous with screen. Several displays may be presented sequentially on the screen, but a program never displays a previous screen.

Dissolve: A transition between two shots, where one shot fades away and simultaneously another shot fades in. Dissolves are done at the lab in the printing phase, but prepared by the negative cutter, who cuts in an overlap of the two shots into the A&B rolls. Labs will only do dissolves in fixed amounts, such as 24 frames, 48 frames, etc.

Distanciation: A technique of Brechtian performance style wherein the actor retains the sense of him/herself as an actor; thus the viewer and actor alike are distanced from the character rather than identifying with it.

Distortion: An undesirable effect that can occur in audio and video signals. In audio, distortion may manifest itself as discordant or harsh sound, or as static. In video, distortion may appear as waves, snow, or incorrect colors.

Distribution Amplifier: A device which splits (distributes) one audio and/or video source to several audio/video device inputs. Typically, distribution amplifiers are used in duplication studios where many tape copies must be generated from one source or in multiple display setups where many monitors must carry the same picture, etc.

Distribution Quality: The level of quality of a program feed from a network to its television stations. MPEC2 @ ML is an example of distribution quality compression.

Distribution: The process of delivering digital media that includes hosting, webcasting, and syndication.

Distributor: The organization responsible for coordinating the distribution of the finished movie to exhibitors, as well as the sale of videos, laserdiscs, and other media versions of movies.

Dither: A computer-aided process of adding dots to a picture to order to smooth the appearance of an image.

Diversity Antennas: A radio-frequency (RF) microphone system using two or more receiving antennas in an attempt to eliminate or reduce multi-path interference.

DIVTC: See Digital Vertical Interval Time Code.

DIVX (Digital Video Express): A short-lived pay-per-viewing-period variation of DVD.

DLT: See Digital Linear Tape.

DMA: See Designated Market Area.

Dockable: Camera or VCR feature whereby the two can work independently or can be joined into a single camcorder unit.

Docudrama: A dramatic production based on factual events or an actual situation.

Documentary: A non-fiction narrative without actors. Typically a documentary is a journalistic record of an event, person, or place. The word "documentary" is rooted in the earliest, anthropological use of film to "document" or record field work. Now, it can mean everything from a classic documentation of life in a remote part of the world to a commissioned work with a clear persuasive goal to a filmmaker's personal exploration of any subject under the sun. Documentaries that are broadcast on public television or by the news departments of commercial television stations are still generally required to be independently funded, as opposed to commissioned by an advocacy organization with a specific agenda. See also Cinema Verité.

Dolby Digital (formerly Dolby AC-3): The approved 5.1 channel (surround-sound) audio standard for ATSC digital television, using approximately 13: 1 compression. Six discrete audio channels are used: Left, Center, Right, Left Rear (or side), Right Rear (or side), and a subwoofer: LFE, "low frequency effects": (considered the ".1" as it is limited in bandwidth).

Dolby Digital /AC-3 Compatible: Dolby Digital provides 6 independent soundtrack channels through the optical output jack. When connected to the AC3 compatible home theater audio set-up, this connection provides Dolby Digital Surround Sound (when Dolby Digital is part of the programming being viewed).

Dolby Laboratories: Founded in 1965, Dolby Laboratories is well known for the technologies it has developed for improving audio sound reproduction, including their noise reduction systems (e.g., Dolby A, B, and C), Dolby Digital (AC-3), Dolby Surround, and more.

Dolby Pro Logic: Dolby Pro Logic (known as Dolby Surround in the theaters) is based on the use of an amplitude-phase matrix. This is a method of encoding four channels of information into two tracks of stereo media and then decoding them back into four channels for playback.

Dolby SR: Spectral Recording. An encoding/decoding noise reduction system developed by Dolby Laboratories and used increasingly in film sound.

Dolly Grip: Crewmember who moves a dolly.

Dolly Shot: Any shot made from a moving dolly. These may also be called tracking or traveling shots.

Dolly Tracks: A set of tracks upon which a camera can be moved on.

Dolly: A camera term referring to when a camera moves in or out of a shot. There are two dolly moves, dolly in and dolly out. A dolly is usually done on wheels or tracks depending on the budget of the film.

Domain: Program Chains (PGC) are classified into four types of domains, including First Play Domain, Video Manager Menu Domain, VTS Menu Domain and Title Domain.

Doorway Dolly: A plywood dolly with four soft tires which is narrow enough to fit through a doorway. It is used to carry a camera on a tripod or for transporting other heavy items.

Dope Sheet: A list of scenes from the script that have already been filmed, or a list of the contents of an exposed reel of film stock. An accurate dope sheet is the responsibility of the assistant cameraman. Also referred to as a Camera Report.

Dot Crawl: Picture degradation resulting from the color modulation scheme used to create composite analog television. Cross-luminance artifacts are caused by a change of the luminance phase from field to field near edges of objects in a scene due to imperfect cancellation of the subcarrier.

Dot Pitch: The distance between screen pixels measured in millimeters. The smaller the number, the better the horizontal resolution.

Dots: Small nets and flags used to control light.

Double Bill: Two movies shown consecutively, typically for a discounted single admission price. Often the movies are sequels or are otherwise related (by genre, eg). See also Feature Presentation, Supporting Feature, and Trailer.

Double Buffering: For video, this means two frame buffers. While buffer 1 is being read, buffer 2 is being written to. When finished, buffer 2 is read out while buffer 1 is being written to.

Double Exposure: A double exposure occurs when (prior to development) an exposed piece of film is reshot with a second image on top of the first. Several exposures can be made, but it still valid to call it a "double" exposure rather than a "triple" or "quadruple" exposure. It is perfectly alright to say "five double exposures," as numerically incongruous as it may sound.

Double Fog Effect Filter: A combination filter which uses soft fog with a heavy low contrast filters. The results in clearer detail than the standard fog filter, while maintaining a dense fog appearance.

Double Headset: Earphones which carry the PL (production line) audio in one ear and program audio in the another.

Double Perf: 16mm film with a row of perforations running along both edges. On the film can this will be indicated by 2R appearing on the label.

Double Reel: In 35mm a double reel is 2 single reels joined together, the maximum size being 2,000 feet. Double reels are labeled 1 ^{A}/B, 2 ^{A}/B etc., to distinguish them from single reels.

Double Re-entry: A switcher which permits the output of one mix/effects bus to be reentered into another mix\effects bus for further video manipulation.

Double System Projector: A projector designed to project a workprint and play a mag track in sync.

Double System: The term double system refers to sound and picture as two separate elements, recorded, edited or projected in sync. 16mm and 35mm use the double system format. A camera photographs the picture and a tape recorder records the sound. In the end, the final print is

Single System, combining sound and picture onto the same piece of print stock.

Double: An actor who stands in for another actor in certain scenes, some of which may involve dangerous circumstances or require special skills (e.g. a stunt double). Sometimes body doubles are used in scenes that call for nudity or intimacy.

Doubler: See Line Doubler

Double-Recorded: Making two recordings of a production in case one experiences a technical problem.

Downconversion: The reconstruction of a video frame or image to have a smaller number of scan lines. This must be done to put an HDTV picture on a standard TV set. See also Sampling.

Downconverter: A circuit used to change a high-frequency signal to a lower frequency.

Downlink: A signal's path from satellite to antenna. It's also the transfer of information from one computer to another, usually through electronic means such as a modem or network. This usually implies transfer from a large computer to a smaller computer.

Download: Process of receiving data from a large computer, digital camcorder, or VCR to a secondary machine such as a PC.

Downloadable Media: A media delivery method where the entire file is transferred to a local machine before playback is allowed.

Downmix: To convert a multichannel audio track into a two-channel stereo track by combining the channels with the Dolby Surround process. All DVD players are required to provide downmixed audio output from Dolby Digital audio tracks.

Downsampling: See Subsampling.

Downstream Keying (DSK): An effect available in some

special effects generators and video mixers in which one video signal is keyed on top of another video signal. The lightest portions of the DSK signal replace the source video leaving the dark areas showing the original video image. Optionally, the DSK signal can be inverted so the dark portions are keyed rather than the lightest portions allowing a solid color to be added to the keyed portions. The DSK input is most commonly a video camera or character generator. The DSK signal must be genlocked to the other signals.

DPI (Dots Per Inch): The number of image-producing dots, to a linear inch. The unit is used to describe the resolution of computer printers.

Draftsman: A person who creates the plans for set construction. See also swing gang, production designer, and art director.

Drag: Amount of friction in the camera head restricting panning and tilting movements.

Dramatic Format: A script format associated with film and teleplays.

Dramatic License: Generally refers to embellishing a true story with elements designed to simplify its structure or increase its dramatic impact.

DRC: See Dynamic Range Compression.

Dress Rehearsal: A camera rehearsal. Final rehearsal with final facilities, sets and wardrobe. Often videotaped.

Drift: Flutter which occurs at random rates.

Driver: Software that provides an operating interface between computer devices. For example, a driver is needed to make a particular video monitor work with a computer and computer operating system.

Drop Field Scrambling: This method is identical to the sync suppression technique for scrambling analog TV channels, except there is no suppression of the horizontal blanking intervals. Sync pulse suppression only takes place during the vertical blanking interval. The descrambling pulses still go out for the horizontal blanking intervals (to fool unauthorized descrambling devices). If a descrambling device is triggering on descrambling pulses only, and does not know that the scrambler is using the drop field scrambling technique, it will try to reinsert the horizontal intervals (which were never suppressed). This is known as double reinsertion, which causes compression of the active video signal. An unauthorized descrambling device creates a washed-out picture and loss of neutral sync during drop field scrambling.

Drop Frame Time Code: A system which electronically skips frames at regular intervals to compensate for the timing difference which exists between black and white and color signals.

Drop In: The process of inserting recorded audio by playing up to a chosen point and switching from playback to record mode.

Drop Out: Loss of a portion of a signal, usually due to a loss of a tape's oxide coating or due to dirt or grease covering a portion of a tape.

Drop Shadow: Shadows, generally electronically produced, that are added around letters.

Drop: Backgrounds of scenery painted on canvas.

Dropped Frames: Missing frames lost during the process of digitizing or capturing video. Dropped frames can be caused by a hard drive incapable of the necessary data transfer rate.

Dry Run: A rehearsal of actors without equipment generally intended to establish blocking.

DSI: See Data Search Information.

DSK: See Downstream Keying.

DSP: See Digital Signal Processor.

DTH (Direct to Home): Official term used by the Federal

Communications Commission industry to refer to the satellite television and broadcasting industries.

DTO: See Discrete Time Oscillator.

DTS: See Digital Theater Systems.

DTS-ES: A version of DTS decoding that is compatible with 6.1-channel Dolby Surround EX. DTS-ES Discrete is a variation of DTS encoding and decoding that carries a discrete rear center channel instead of a matrixed channel.

DTV: See Digital Television.

Dual Channel: A SCSI card design which employs separate interface chips for two different drives, intended to almost double the effective disk I/O throughput.

Dual Feed: The use of two transponders to broadcast programs. Typically one feed for the Eastern and Central time zones (called the eastern feed) and the second feed for the Pacific time zone (called the western feed). One of the advantages of a dual feed means that a network's prime time programming will air 8-11PM on both the east and west coast. All broadcast and pay cable networks (measured by Nielsen) use a dual feed. Several ad supported cable networks also use a dual feed including The Family Channel.

Dual-redundancy: The use of two identical condenser microphones on a subject, one intended as a backup in case the primary microphone goes out.

Dub Stage: Term generally used in California but to refer to the room where the final audio mix is made for a program or film. It might also be known as a mix stage.

Dubbing: The recording of dialogue in a sound studio, after the footage is shot, where the actors watch the film and match the lip movements.

Duct Tape: A shiny adhesive tape designed for holding metal heating and cooling ducts, but also commonly used as a substitute for gaffers tape, a general purpose tape used in television and film.

Dulling Spray: A wax-based aerosol spray used on shiny surfaces to reduce reflections.

Dupe: A dupe is a positive copy of a positive. A dupe can also be a negative copy of a negative. A dupe is a print made in the reversal process. It can sometimes be clearer to call something a dupe, because to simply say "positive print" you could just mean a positive copy of a negative, which would not be a dupe.

Duplication House: A company that duplicates videocassettes.

Duplication: The reproduction of media. Generally refers to producing discs in small quantities, as opposed to large-scale replication.

Dutch Angle: A tilted camera angle that shows images obliquely slanted to the frame's vertical axis. Also called Oblique Angle.

Dutch Tilt: A composition with the camera viewing the scene at a diagonal. Same as a canted angle. Some nice examples can be seen in Carol Reed's "The Third Man."

Duvetyne: A heavy black cloth, treated with fire proofing material, which is used for blacking out windows, making teasers, hiding cables, and hundreds of other uses.

DV: See Digital Video.

Dv_export: An export mode in Adobe Premiere that enables digital video to be exported through a capture card.

DV25: The most common form of DV compression. DV25 uses a fixed data rate of 25 megabits per second.

DVB: See Digital Video Broadcast.

DVC (Digital Video Cassette): A digital tape recording format using approximately 5:1 compression to produce near-Betacam quality on a very small cassette. Originated as a consumer product, but being used professionally as exemplified by Panasonic's variation, DVC-Pro, and Sony's variation DVCam.

DVC Pro: Panasonic professional videocassette format based on DVC but using a wider track and faster tape speed for higher quality.

DVCAM: Sony's proprietary version of DV.

DVD (Digital Versatile Disc): Digital Versatile Discs resemble audio CDs in appearance, but have a much higher storage capacity. Hence, they can store rich digital media such as video in addition to audio and computer software. DVD was once called "Digital Video Disc" but the name change reflects its wider uses. As a video medium, DVD offers full length feature films to be stored with exceptional picture quality accompanied by high end digital sound, such as Dolby Digital and DTS. Thanks to the huge capacity of discs, DVD movies are often sold with extra features such as the option to view the movie in widescreen or fullscreen, or the option to listen to the movie or director commentaries.

DVD Forum: An international association of hardware and media manufacturers, software firms and other users of Digital Versatile Discs, created for the purpose of exchanging and disseminating ideas and information about the DVD Format.

DVD Multi: DVD Multi is a logo program that promotes compatibility with DVD-RAM and DVD-RW. It is not a drive, but defines a testing methodology which, when passed, ensures the drive product can in fact read RAM and -RW. It puts the emphasis for compatibility on the reader, not the writer.

DVDA: DVD Association, a non-profit industry trade association representing DVD authors, producers, and vendors throughout the world.

DVD-Audio (DVD-A): The audio-only format of DVD. Primarily uses PCM audio with MLP encoding, along with an optional subset of DVD-Video features.

DVDO: Trademark of Silicon Image Corp. for its line of de-interlacers and de-interlacing chip sets; also name of company that made same and that was acquired by Silicon Image Corp.

DVD-RAM (DVD Random Access Memory): A rewritable DVD disc endorsed by Panasonic, Hitachi and Toshiba. It is a cartridge-based, and more recently, bare disc technology for data recording and playback. The first DVD-RAM drives were introduced in Spring 1998 and had a capacity of 2.6GB (single sided) or 5.2GB (double sided). DVD-RAM Version 2 discs with 4.38GB arrived in late 1999, and double-sided 9.4GB discs in 2000. DVD-RAM drives typically read DVD-Video, DVD-ROM, and CD media.

DVD-ROM: The base format of DVD. ROM stands for read-only memory, referring to the fact that standard DVD-ROM and DVD-Video discs can't be recorded on. A DVD-ROM can store essentially any form of digital data.

DVD-RW (DVD ReWritable): A rewritable DVD format, introduced by Pioneer, that is similar to DVD+RW. It has a read-write capacity of 4.38 GB.

DVD-Video (DVD-V): A standard for storing and reproducing audio and video on DVD-ROM discs, based on MPEG video, Dolby Digital and MPEG audio, and other proprietary data formats.

DVE: See Digital Video Effects.

DVI: See Digital Visual Interface.

DVNR: See Digital Video Noise Reduction.

DVS (Descriptive Video Services): Descriptive narration of video for blind or sight-impaired viewers.

Dye Polymer: The chemical used in DVD-R and CD-R media that darkens when heated by a high-power laser.

Dye-sublimation: Optical disc recording technology that uses a high-powered laser to burn readable marks into a layer of organic dye. Other recording formats include magneto-optical and phase-change.

Dynamic Composition: Elements of composition related

to the moving image and to the interrelationship between scenes.

Dynamic Contrast Control: Camera circuit that increases contrast ratio beyond 30:1, allowing overly bright and very dark areas to be successfully reproduced.

Dynamic Distortion: Alteration of volume range of a sound when it is transmitted.

Dynamic Editing: Editing with primary regard to building an effect rather than presenting scenes in a logical, linear fashion.

Dynamic Limiter: An audio circuit which restricts the dynamic range of audio.

Dynamic Microphone: A microphone which uses a magnet moving in a coil of wire to generate an electrical signal.

Dynamic Montage: Editing intended to evoke strong emotional reactions. See Russian montage.

Dynamic Q: A video data-rate control technique which effectively changes the video compression ratio on a frame by frame basis to constrain the video data-rate to within the available system bandwidth. Dynamic Q therefore helps minimize dropped frames.

Dynamic Range Compression: A technique of reducing the range between loud and soft sounds in order to make dialogue more audible, especially when listening at low volume levels. Used in the downmix process of multichannel Dolby Digital soundtracks.

Dynamic Range: A range of sounds from soft to loud. A measurement of the limits of microphones, recording and playback machines, and other audio equipment.

Dynamic Tracking: A video head's ability to find and follow an adjacent track by bending back and forth. Dynamic tracking offers true freeze-frame instead of freeze-field, variable-speed playback, and playing in reverse.

E

E to E (Electronics to Electronics): The situation where the input to a video or audio machine appears on its output. E to E is used in linear editing to perform a preview edit. The record machine will play back the video and audio from the record tape and at the in point go into to E to E mode showing the incoming source but not recording the result. Produces some distortion and timing errors, therefore signals should be used direct and not via E to E machines.

E-1 Circuit: An E-1 Circuit (2,000,000 bits per second) is the European equivalent of a T-1.

Ear: To put a flag up on the side of a lighting unit to block light. Better known as a "sider."

Early Fringe: The television time period preceding prime time access, (usually 4PM to 7PM, except in the Central Time Zone where it extends from 3PM to 6PM). See Prime Time.

Eastman Color Negative (ECN): A developed negative.

EAU: See European Broadcast Union.

EBCDIC: (pronounced eb-si-dick) The abbreviation for Extended Binary Coded Decimal Interchange Code, EBCDIC is a scheme for representing the internal coding structure for most IBM systems. All numbers, letters, and most special characters are represented. Usually a television station uses this term only when specifying the print belt that is to be used on a line printer.

EBR: See Electron Beam Recording.

EC Camera (Electronic Cinematography Camera): A high-quality video camera designed to closely resemble a film camera, both in operation and technical quality.

ECC Constraint Length: The number of sectors that are

interleaved to combat bursty error characteristics of discs. 16 sectors are interleaved in DVD. Interleaving takes advantage of typical disc defects such as scratch marks by spreading the error over a larger data area, thereby increasing the chance that the error correction codes can conceal the error.

ECC: See Error Correction Code.

Echo: A sound wave that has been reflected and returned with sufficient magnitude and delay to be perceived as a wave distinct from that which was initially transmitted.

ECN: See Eastman Color Negative.

ECU: See Extreme Close Up.

Ed-Beta (Extended Definition Betamax): A high-quality consumer/professional videocassette format developed by Sony.

EDC: See Error Detection Code.

Edge Enhancement: When films are transferred to video in preparation for DVD encoding, they are commonly run through digital processes that attempt to clean up the picture. These processes include noise reduction (DVNR) and image enhancement. Enhancement increases contrast (similar to the effect of the "sharpen" or "unsharp mask" filters in PhotoShop), but can tend to overdo areas of transition between light and dark or different colors, causing a "chiseled" look or a ringing effect like the haloes you see around street lights when driving in the rain. Video noise reduction is a good thing, when done well, since it can remove scratches, spots, and other defects from the original film.

Edge Fog: Exposure along the edge of the film from raw light, in most cases from a lightleak, due to the camera door not being taped. Edge Fog can sometimes be visible in the frame or sometimes outside of the frame effecting the clarity of the latent edge numbers.

Edge Numbers: Small numbers running along the edge of

the film, in between the perf in 16mm, and just to the far side of them in 35mm. The are photographed onto the film in its manufacture, and are there to aid the negative cutter in lining up shots in the process of conforming the negative. They are sometimes called latent edge numbers to distinguish them from inked-on code numbers. Code Numbers are sometimes called edge numbers.

Edge Track: A standard position for the placement of the audio on a single perforation magnetic film.

EDH: See Error Detection Handling.

Edison Plug: An ordinary household plug with two flat blades and a ground pin.

Edit Control: A connection on a VCR or camcorder which allows direct communication with external edit control devices. (e.g., LANC (Control-L) and NEW (Panasonic) 5-pin).

Edit Decision List (EDL): Complete list of time code numbers for each shot and sound used in the off-line edit master. These time code numbers are used to create the final on-line edit master.

Edit Master: Video industry term for the tape containing the finished (edited) program.

Edit Points: Also known as "edit in" and "edit out." The beginning and end points of an edit when a video program or soundtrack is being assembled.

Edit Recorder: The destination VTR that records the source tapes onto the edited master during the editing session.

Edit: The cutting and arranging of shots. In the different stages, or at the completion of editing the edited film itself can be referred to as "the cut" or "the edit."

Editech: The first electronic editing system for videotape: invented and marketed by Ampex.

Edited Master: The final tape created during the editing process.

Editing Bench: A workbench with rewinds attached, and sometimes a built-in light table in the center.

Editing Bin: See Trim Bin.

Editing Block: A metal editing device which holds two pieces of audio tape in place for cutting and splicing.

Editing Pace: The number of edits per unit of time.

Editing Workstation: A complete computer-controlled, desktop system used to edit audio and/or video productions.

Editing: The work of selecting and joining together shots to create a finished film.

Editor: A person who performs editing (in consultation with the director) on a movie. This term usually refers to someone who does visual editing. See also Motion Picture Editors Guild.

Editorial Assistant: A videotape editor's assistant.

Editorial Sync: A set of sync marks on picture and sound that line up at the same frame, as opposed to Printer's Sync, where the picture and sound are displaced. Sometimes it is useful to label a sync mark ES to know that it is an Editorial Sync mark.

EDL: See Edit Decision List.

EDO (Extended Data Out): A high performance DRAM memory architecture which effectively doubles the memory bandwidth. EDO memory is recommended for use with Plum.

EDS (Enhanced Data Services): Additional information in NTSC line such as a time signal.

EDTV: See Enhanced Definition Television.

Effective Output Level: The microphone sensitivity rating defined as the ratio in dB of the power available relative to sound pressure.

Effective Reach: The percentage of the target audience that is exposed to the advertising schedule a sufficient number of times to produce a positive change in awareness, attitude or purchasing action. It is based on the concept that exposure below the effective frequency has little or no value.

Effects Bank: A row of buttons on a switcher that can be used for fades, wipes, or special effects.

Effects Monitor: A video monitor where video effects are set up and checked before use.

Effects Theory: A type of communication theory (e.g., hypodermic needle concept) which proposes that, because viewers are passive, television directly affects them.

EFM (Eight-to-fourteen Modulation): This low-level and very critical channel coding technique maximizes pit sizes on the disc by reducing frequent transitions from 0 to 1 or 1 to 0. CD represents 1's as Land-pit transitions along the track. The 8/14 code maps 8 user data bits into 14 channel bits in order to avoid single 1's and 0's, which would otherwise require replication to reproduce extremely small artifacts on the disc. In the 1982 compact disc standard (IEC 908 standard), 3 merge bits are added to the 14 bit block to further eliminate 1-0 or 0-1 transitions between adjacent 8/14 blocks.

EFP: See Electronic Field Production.

EGA (Enhanced Graphics Adapter): A relatively high-resolution computer color graphics adapter for PCs.

EI: See Exposure Index.

EIA RS-170A: The timing specification standard for NTSC broadcast video equipment.

EIA/IS-702: NTSC Copy Generation Management System: Analog (CGMS-A). This adds copy protection capabilities to NTSC video by extending the EIA-608 standard to control the Macrovision anti-copy process.

EIA: See Electronic Industries Association.

EIA-516: United States teletext standard, also called NABTS.

EIA-608: United States closed captioning and extended data services (XDS) standard.

EIA-708: United States DTV closed captioning standard.

EIA-744: NTSC "v-chip" operation. This adds content filtering capabilities to NTSC video by extending the EIA-608 standard.

EIA-761: Specifies how to convert QAM to 8-VSB, with support for OSD (on screen displays).

EIA-762: Specifies how to convert QAM to 8-VSB, with support for OSD (on screen displays).

EIA-766: United States HDTV content advisory standard.

EIA-770: This specification consists of three parts (EIA-770.1, EIA-770.2, and EIA-770.3).

EIA-775: This standard defines a specification for a baseband digital interface to a DTV.

EIA-J CPR-1204: This EIA-J recommendation specifies another widescreen signaling (WSS) standard for NTSC video signals. WSS may be present on 20 and 283.

E-IDE (Enhanced Integrated Drive Electronics): Extensions to the IDE standard providing faster data transfer and allowing access to larger drives, including CD-ROM and tape drives, using ATAPI. E-IDE was adopted as a standard by ANSI in 1994. ANSI calls it Advanced Technology Attachment-2 (ATA-2) or Fast ATA.

EISA (Extended Industry Standard Architecture): An interim base architecture between ISA and PCI, which is basically dead.

Electret Microphone: A microphone that uses a condenser-type sound sensing element.

Electrical Department: Section of a production's crew responsible for lighting and other electrical matters during filming.

Electrician: Grip that specializes in the equipment of the electrical department.

Electron Beam Recording (EBR): A videotape-to-film transfer process in which the three primary colors of a color picture are electronically scanned onto a color (or onto three black and white) films.

Electron Gun: A mechanical device, located in the rear of a television's picture tube, which fires an electron beam at the pixels, scanning line-by-line across the lines of the television image, causing the pixels to glow and create the television image.

Electronic Beam Recording: A technique for converting television images to film using direct stimulation of film emulsion by an electron beam.

Electronic Cinematography: Videography. Using video cameras to do production, especially single-camera production.

Electronic Effects: Special effects (including fades, dissolves and keying) created on video using an analog special effects generator.

Electronic Field Production (EFP): Production of a television program or program segment by using portable video cameras, VTRs, and sound equipment outside the studios. EFP is characterized by generally higher production quality than ENG.

Electronic Graphics: Titles, credits, drawings, etc., produced electronically.

Electronic Industries Association (EIA): An organization that determines audio and video standards.

Electronic Palette: Peripheral device used to input commands into a computer. An electronic pencil is commonly used with the palette to draw shapes and select drawing and paint functions.

Electronic Pencil: Pencil-like, hand-to-computer device used to input commands and make drawings. Commonly used with an electronic palette.

Electronic Pin Register (EPR): Stabilizes the film transport of a telecine. Reduces ride (vertical movement) and weave (horizontal movement). Operates in real time. See also Steady Gate.

Electronic Programming Guide (EPG): An application that provides an on-screen listing of all programming and content that an interactive television service subscriber or digital television viewer has available.

Electronic Still Store (ESS): An electronic device that can capture a full frame of video and store it in digital form.

Electronic Test Pattern (ETP): A test pattern generated within electronic circuitry used for aligning equipment.

Electronic Viewfinder: A small picture tube (CRT) or liquid crystal display (LCD) built into a video camera enabling the operator to see what the camera is photographing.

Electrostatic Microphone: A type of microphone that uses a sound-sensing element similar in principle to that of a condenser microphone.

Elementary Stream (MPEG): A coded MPEG bitstream for one program of audio or video or one data stream.

Elevation: How high a satellite is from the horizon.

Ellipsoidal Spotlight: A spotlight producing a focused, sharply-defined beam of light.

Embedded Audio: Method of transmitting a digital video signal together with accompanying (multi-channel) sound by way of insertion of audio data packets into the ancillary data capacity of the serial digital video stream. (SMPTE standard 272M).

Emotional Memory: Technique of method acting wherein the actor draws upon memories of previous emotions that match the emotions of the character.

Emulate: To test the function of a DVD disc on a computer after formatting a complete disc image.

Emulsion Leader: Unlike plastic leader, emulsion leader can be cement spliced.

Emulsion Speed: The photosensitivity of an emulsion, usually expressed as an index number based on the film manufacturer's recommendations. See Slow-speed Film and High-speed Film.

Emulsion Type: A film's emulsion type refers to the composition of its emulsion, whether it was manufactured to be fast, slow, grainy, fine-grained, colorful, pastel, black and white or color, daylight balanced, tungsten balanced, etc. The emulsion type is represented by a number. For Kodak it is a series of four numbers, such as 7248. The "72" always stands for 16mm camera stock, and the same emulsion type is found in 35mm as 5248, "52" being the designation of 35mm. Fuji uses a system where the film's emulsion type is a little more telling, such as 250D, which is daylight balanced film with an Exposure Index of 250. When picking out a stock to use the film speed, and in the case of color film, whether the film is daylight or tungsten, are the primary reasons for choosing a certain emulsion type. Allowances might also be made to achieve a certain look, as in using Kodak Vision, or Fuji film. Several different emulsion types are usually used on a project, fast for night scenes, slow for daylight scenes, etc.

Emulsion: The thin layer of silver attached to the base which, when exposed and developed, creates the film image through the areas of silver, which block light, and the clear areas which allow light to pass through.

Encode: The process of combining analog or digital video signals, e.g., red, green and blue, into one composite signal. In cultural studies, encode means the creation of meaning within a text by a cultural institution such as the television industry. Readers/viewers may decode these preferred meanings when exposed to texts, or they may take a position opposing them.

Encryption: The process of coding data so that a specific code or key is required to restore the original data, used to make transmissions secure from unauthorized reception.

Endoscopic Television: Obtaining television images inside a living body, generally by inserting fiber optic probes.

ENG Coordinator: The individual under the assignment editor and news producer responsible for integrating the resources of the technical and news personnel in a TV news room.

Enhanced Data Services: See EDS.

Enhanced Definition Television (EDTV): A television capable of displaying at least 480 progressive active scan lines. No aspect ratio is specified. For the ATSC system, typical EDTV (luminance) resolutions are:

Active Resolution	Aspect Ratio
640 x 480	4:3
720 x 360	16:9
720 x 480	4:3

Enhanced for 16: 9: For DVD, refers to a program where the entire 720 by 480 pixel video frame represents a 16: 9 aspect ratio image. Whereas for a "standard widescreen" program, only the middle 720 by 360 pixels represent a 16: 9 image with the topmost and bottommost pixels unused. To show the full resolution, the enhanced program must be viewed on a TV or monitor with an adjustable image height control or a 16: 9 aspect ratio screen. The exact number of pixels of picture height excluding letterbox bars will vary depending on the exact aspect ratio of the image, although the enhanced version will still have better resolution than a standard letterbox edition.

Enhancing: Improving a video image by boosting the high frequency content lost during recording. There are several types of enhancement. The most common accentuates edges between light and dark images.

Entropy Coding: Variable-length, lossless coding of a digital signal to reduce redundancy. MPEG-2, DTS and

Dolby Digital apply entropy coding after the quantization step. MLP also uses entropy coding.

Envelope: The shape of the graph as amplitude is plotted against time. A sound's envelope includes its attack, decay, sustain and release (ADSR).

Environmental Sound: General low-level sound coming from the action of a film, which can either synchronous or non-synchronous.

EPG: See Electronic Programming Guide.

Epic Theater: Brechtian theory of theatrical presentation in which the viewer is alienated from the character.

Epic: A film with large dramatic scope or that required an immense production.

EPR: See Electronic Pin Register.

Equal Time: The FCC's Equal Opportunities Rule. If a television station gives or sells airtime to one candidate for public office, it must offer equivalent time to other candidates. News programs are exempt from this rule.

Equalization Pulses: These are two groups of pulses, one that occurs before the serrated vertical sync and another group that occurs after. These pulses happen at twice the normal horizontal scan rate. They exist to ensure correct 2:1 interlacing in early televisions.

Equalization: The alteration of sound frequencies for a specific purpose, such as to remove "noise" frequencies or to improve speech clarity.

Equivalent Noise: A microphone in a completely silent room still generates some residual noise. This noise can be measured and can be computed. That computation is the "Equivalent Noise."

Erase Head: A degaussing magnetic field used in a video or audio recorder to eliminate previously recorded material on the tape.

Erase: Electronically wiping clean or degaussing a video or audio tape.

Error Concealment: A technique used when error correction fails. Erroneous data is replaced by data synthesized from surrounding pixels.

Error Correction: Digital method of checking if digital data was transmitted or recorded correctly. If an error is found, the data is resent.

Error Detection and Handling (EDH): SMPTE RP-165 for recognizing errors in the serial digital signal.

Error Detection Code (EDC): A 32-bit (4 byte) CRC-like code appended at the end of a DVD data sector.

Errors of Continuity: Disruptions in the flow of a scene, such as a failure to match action or the placement of props across shots.

ES: Editorial Sync.

ESS: See Electronic Still Store.

Essential Area: See Safe Action Area.

Establishing Shot: A shot, normally taken from a great distance or from a "bird's eye view," that establishes where the action is about to occur.

Estar Base: A brand name for Polyester Base.

Ethernet: Ethernet is a local area network transport protocol that is one of the most pervasive in the networking industry. It offers a 10 megabit (10,000,000 bits per second) speed for data throughput.

Ethnographic Film: An anthropological film that records and perhaps comments on an ethnic group and its culture.

ETP: See Electronic Test Pattern.

ETR 154: This specification defines the basic MPEG audio and video parameters for DVB applications.

ETS 300163: This specification defines NICAM 728 digital audio for PAL.

ETS 300231: This specification defines information sent during the vertical blanking interval using PAL teletext (ETS 300706) to control VCRs in Europe (PDC).

ETS 300294: Defines the widescreen signaling (WSS) information for PAL video signals. For (B, D, G, H, I) PAL systems, WSS may be present on line 23.

ETS 300421: This is the DVB-S specification.

ETS 300429: This is the DVB-C specification.

ETS 300706: This is the enhanced PAL teletext specification.

ETS 300707: This specification covers Electronic Program Guides (EPG) sent using PAL teletext (ETS 300706).

ETS 300708: This specification defines data transmission using PAL teletext (ETS 300706).

ETS 300731: Defines the PALplus standard, allowing the transmission of 16**:** 9 programs over normal PAL transmission systems.

ETS 300732: Defines the ghost cancellation reference (GCR) signal for PAL.

ETS 300743: This is the DVB subtitling specification.

ETS 300744: This is the DVB-T specification.

ETSI: European Telecommunications Standards Institute.

European Broadcast Union (EAU): An organization of European broadcasters.

Event Video: The videotaping of special events, such as a weddings, graduations, etc.

Event: A generic term for a program or spot that occurs on-air. Each item that has to be broadcast is an event.

Exceptional Quality: An element of newsworthiness

which distinguishes a story or plot from the commonplace and captures and holds viewer attention.

Exciter Lamp: A special lamp in the projector used for the playback of Optical Sound. The projector reads the track by passing it between the exciter lamp, a light-sensitive photo-electric cell.

Exclusivity: The advertiser pays a premium or a major commitment guaranteeing all other competing products will be prohibited from a program or cable network.

Executive Producer: A producer who is not involved in any technical aspects of the filmmaking process, but who is still responsible for the overall production. Typically an executive producer handles business and legal issues.

Exhibition: The act of screening the feature film to the public.

Exhibitor: An organization which represents cinemas.

Existing Light: The normal, prevailing light at a location.

Expander: An audio device that increases the dynamic range of an audio signal.

Experimental Film: See Underground Film.

Expose: An investigative documentary that reveals, often in shocking ways, discreditable information or events.

Expository Mode: Mode of television that presents an argument about the historical world; the "facts" of that world are assertively or even aggressively selected and organized and presented to the viewer in a direct address.

Exposure Index (EI): This is the sensitivity to light of a particular type of film. It is the specific number used to measure Film Speed. The film will list an E.I. number on the box or the film can as the film speed. It is the same as A.S.A. and I.S.O. on the light meter.

Expressionism A style of filmmaking which distorts physical reality in some way in order to "express" strong feelings about it. Typical expressionistic techniques include the use of distorting lenses, extreme camera angles, bizarre lighting and sound effects, and fragmented editing. See realism.

Extended Kell Factor: See Kell Factor.

Extension Tubes: These are a handy way to turn any long lens into a macro lens for ultra-close shooting. They are hollow metal tubes that are mounted between the camera and the lens. Typically they come in a set of different lengths which can be combined. It is a good idea to open up the lens a little when using an extension tube, as a little light is lost. It should be noted that they do not work when used with wide lenses.

Extent: For the volume structure and the ISO 9660 file structure, an extent is defined as a set of logical sectors, the logical sector numbers of which form a continuous ascending sequence. The address, or location, of an extent is the number of the first logical sector in the sequence. For the UDF file structure an extent is defined as a set of logical blocks, the logical block numbers of which form a continuous ascending sequence. The address, or location, of an extent is the number of the first logical block in the sequence.

Exterior Scenes: Scenes set outdoors, often in particular location settings.

External Key: As opposed to an internal key, a key which depends on three video sources: background video, a video source to determine the shape of the area to be keyed out of the background video, and a video source to be placed in the keyed out area.

External Sync: As opposed to internal sync, video synchronizing pulses from an external source, such as a sync generator. Used to synchronize multiple pieces of equipment.

Extra: A person who appears in a movie where a non-specific, non-speaking character is required, usually as part of a crowd or in the background of a scene. Extras are often recruited from wherever they are available. Contrast with non-speaking role.

Extreme Close-Up (XCU or ECU): A framing that presents a view closer than a conventional close-up: e.g., a shot of an eye that fills the entire screen.

Extreme Long Shot (XLS): A framing that presents a distant view of an object or person: e.g., an aerial shot of a car on a street.

Eye Light: A small light used in close-ups to illuminate the upper facial area.

Eye Pattern: A digital wave form that can be used to evaluate channel performance.

Eyeline Match: The matching of eyelines between two or more characters. For example, if Sam looks to the right in shot A, Jean will look to the left in shot B. This establishes a relationship of proximity and continuity.

Eyeline: The direction an actor is looking. To maintain continuity during the intercutting of single-camera, film-style scenes, shots of people talking to each other need to have consistent eyelines.

F

F/X: Special audio or video effects.

Facilities Request Form: A form listing all technical facilities needed for a production.

Fade In: Increasing an audio or video from no signal to maximum signal. Often follows a fade out to indicate major divisions within a production. Also the first words typed in a script. Literally means "to begin."

Fade Out: Decreasing an audio or video from maximum signal to zero. Often indicates the end of a segment. Also the last words in a script which means "the end."

Fade: An optical effect in which the screen gradually brightens as a shot appears (fade in) or gradually darkens as a shot goes to black or another blank color (fade out). Sound also fades in or out when the soundtrack gradually changes from silence to sound, or vice versa.

Fader Bars: A pair of levers on the switcher that allow for gradual changes video levels. They can be used to produce superimpositions, dissolves, fades, keys, or wipes at different speeds.

Fader: A volume control or potentiometer used to control the amplitude of an electrical signal.

Fairness Doctrine: The rule repealed by the FCC in 1987 that required broadcasters to devote airtime to important controversial issues and to air contrasting views on those issues.

False Light: Distorting the truth of a situation through such things as wrongful association, or unduly emphasizing negative attributes.

Fantasy Set: Television setting which is abstract or bears only limited resemblance to a realistic setting.

Fast & Wide: The 16-bit, 10 MHz SCSI standard which supports transfer rates up to 20 MBytes/sec.

Fast Film: Film stock that is highly sensitive to light, usually with an exposure index of 100 or higher. Also called Fast-speed Film.

Fast Lens: A lens that transmits a great amount of light. Can be used in low lighting conditions.

Fast Motion: A shot in which time appears to move more quickly than normal. The process is commonly achieved by either deleting select frames (called "skip frames") or by undercranking. Also called Skip Frame.

Fast: The camera assistant's motto. Everything he/she must do before each shot (Focus, Aperture, Shutter, Tach).

Father: The metal master disc formed by electroplating the glass master. The father disc is used to make mother discs, from which multiple stampers (sons) can be made.

Fault: See Bug.

Fay: A 650 watt PAR light with daylight balance dichroic fliter.

FCC: The abbreviation for the Federal Communications Commission established in 1934 that is the regulatory agency governing the broadcasting industry in the United States. The FCC reports to Congress and assigns broadcasting frequencies, licenses television stations, and oversees interstate communications. The FCC is roughly the American equivalent to the CRTC.

Feather: Moving a "flag" closer to or further away from a light source that it is in front of will feather (soften/harden) the shadow on the surface upon which the light falls.

Feature Film: A movie of usually about 90-120 minutes in length on one particular topic.

Feature Presentation: The main or advertised movie during a screening. Also called Main Attraction.

Federal Communications Commission: See FCC.

Feed Horn: A device which collects the signals at the focus of the satellite dish and channels them to the LNB.

Feed Lines: Lines of dialogue which are read outside camera range for the benefit of an 'on camera' or 'on microphone' actor or voice over artist.

Feed: Signal transmission from a specific source. Examples would be a network feed and a remote feed.

Feedback: Sound regeneration caused by a microphone picking up the output from its own speakers. Commonly results in a ringing sound or a high-pitched squeal.

Feeders: Microwave trucks that relay ENG or EFP signals from the field back to the television station.

Feminist Criticism: Film analysis and criticism from a feminist perspective, concerned primarily with the social and political implications of how women are depicted in films.

Femme Fatale: Literally: "Deadly Lady"; a slang term used to describe a character in a movie.

Festival: An event at which films can often premiere. Festivals can be used as by studios to show their wares and sell rights to distributors, or to officially mark a movie's release so as to make it eligible for award ceremonies with hard deadlines that can't be met if they waited for a general release. Some festivals are competitive, giving awards from a jury or selected by the audiences.

Fiber Optics: Thin glass filaments within a jacket or bundle that optically transmits images or signals in the form of light over distances, with extremely low high-bandwidth quality losses.

Fiction Film: Any film that employs invented plot or characters; usually called narrative film.

Field frequency: The number of fields per second. NTSC field frequency is 60-per second PAL frequency is 50 per second.

Field Segmentation: The process of splitting a progressive image frame into two parts (fields) that are then transported and manipulated as two fields of one frame.

Field: A set of alternating scan lines in an interlaced video picture. A frame is made of a top (odd) field and a bottom (even) field.

FIFO: First In First Out, the basic memory used in conjunction with the PCI bus to buffer video and audio data prior to transfer over the bus.

Figure-eight Response: See Bidirectional.

File Extension: Computer name suffix. Last part of a computer file name that comes after the dot; typically .exe, jpg, bmp, etc.

File Transfer Protocol (FTP): The common procedure used for downloading and uploading files over the Internet.

File: A collection of data stored on a disc, usually in groups of sectors.

Fileserver: A central computer that controls network functions and manages shared files.

Fill Leader: Film leader used to fill in the blanks in picture workprint.

Fill Light: A weak light, used to fill in, soften, or eliminate shadows cast by the key light.

Fill Pattern: Solid color or design provided used to fill in, or give texture and pattern to, elements of a chart or a drawing.

Fill: Program material used if the regular program runs short.

Filler: Scrap film most often used to keep a soundtrack running the same length as the picture, even though there is just silence. When used this way in can also be called sound fill. Filler is usually a print with the emulsion scraped off the center all the way along, perhaps to prevent bootlegging, but also useful in that a mark can be seen on both sides through this wide scratch.

Film Base: The flexible, usually transparent support on which photographic emulsions and magnetic coatings are carried.

Film Buyer: A person who arranges to purchase films from an distributor on behalf of an exhibitor.

Film Cement: A liquid that is actually not a glue, but a chemical that melts and fuses two pieces of film together.

Film Chain: Projectors, multiplexers and cameras, connected for the purpose of transferring film to video.

Film Criticism: The analysis and evaluation of films, often according to specific aesthetic or philosophical theories.

Film Grain: The tiny particles of light-sensitive material on film stock that record images. Finer grains give higher image quality, but coarser grains allow a faster shutter speed. Graininess is an artifact which results from the use of coarse grains, and gives images a slight mosaic appearance.

Film Magazines: A reel of film stock ready for use in a camera. The clapper-loader is responsible for inserting these into a camera.

Film Mode: Mode of vertical-temporal video signal processing which assumes that odd and even fields of the TV frame contain information from the same time moment, typical of film.

Film Noir: In French, literally, "black film." A type of film, mainly produced in Hollywood during the 1940s and 1950s, which depicts "dark" themes, like crime and corruption in urban settings, in a visual style that features night scenes and low-key lighting.

Film Notcher: A device used to punch out a small portion of the edge of a piece of film in order to permit electrical contacts on a printer to come together and thus activate light changes.

Film Plane: The film plane is the plane of depth from the lens of the film, behind the gate, in the camera. It is also the point from where the distances on the focusing ring should be measured from, and is indicated on the outside of the camera with a little symbol that looks like the planet Saturn turned on its side.

Film Printing: Process of transferring images from a negative print to a print.

Film Speed: The sensitivity to light for proper exposure of a given film stock. This is primarily a result of the size of the silver halides in the emulsion, the larger the grain, the less light is needed for exposure. Film stocks are generally spoken of as being fast or slow, a fast film having large grains and needing less light, a slow film having smaller grain and needing more light.

Film Stock: The physical medium on which photographic images are recorded. See also Film Grain and Film Stock.

Film Weave: The frame-to-frame misalignment of film in the film gate causes originally static objects to weave on the screen. The problem is most pronounced with old or damaged film where the perforations have been worn. The effect can be removed by stabilization. See also Stabilization and Tracking.

Filmmakers: A collective term used to refer to people who have a significant degree of control over the creation of a movie: directors, producers, screenwriters, and editors.

Filmography: A bibliographic listing of a film artist's body of work.

Film-style Video Production: Single camera video production. Video shot out of sequence with one camera and then assembled into a complete show in postproduction. Typically, multiple takes are made of each scene. This contrasts with switch-feed video productions where multiple cameras are used and the production is edited on a video switcher as it is shot.

Filter: A plastic or glass material placed over or behind a camera lens to alter the light in some way. Primarily used to change color temperature. Also, a device that allows certain parts of an electronic signal to pass while stopping others.

Filter Frame: Holder used on the front of a light or a camera which holds colored gels, filters, diffusers, etc.

Filter Wheel: A holder containing multiple filters that can be rotated into place behind the camera lens.

Filter: A piece of glass or plastic fitted in front of the lens to control the color or quality of light entering the camera.

Final Cut: The finished edit of a film, approved by the director and the producer. This is what the audience sees.

Financial Interest and Syndication Rules: The FCC regulations forbidding TV networks from owning interest in or syndicating most programming that they carry.

Fine Grain: A type of film stock in which the grain is smaller, resulting in a higher image definition.

Fingers: Small flags used to control light.

Finishing: The process of completing a program after the edit decisions have been made. Finishing might include any or all of the following: adding DVE moves, repairing dropouts, adding titles, color correction, building effects, including graphics and even changing the edit decisions.

FIN-SYN: See Financial Interest and Syndication Rules.

Fire Wire: A low-cost digital interface originated by Apple Computer and further developed by engineers and adopted by CEMA. It can transport data at 100, 200, or 400 Mbps. This is widely viewed as one key solution to connect digital-related TV components with each other. Also known as IEEE-1394.

First Draft: As set forth in the Writers Guild of America Minimum Basic Agreement, a first complete draft of any script in continuity form, including dialogue.

First Generation: Either the first print from a film original or a magnetic tape original.

First In First Out: See FIFO.

First Play PGC: This Program Chain (PGC) is described in the Video Manager Information table, and has no corresponding video objects (VOB). The First Play PGC is executed at initial access, e.g. just after disc loading.

First Run Syndication: The national distribution of programs produced for initial release into syndication rather than airing on a network first.

First Unit: The group or crew that performs the acquisition of the primary elements of a media program. A first unit is minimally comprised of the principal actors, director and producers. First unit is also the first production unit that is assembled to photograph scenes to be used in a movie or television program.

First-person Shot: See Point-of-view Shot.

Fisheye Lens: An extreme wide-angle lens that distorts the image so that straight lines appear rounded at the edges of the frame.

Fishpole: A hand-held microphone suspension pole or device.

Five-field Sequence: The relationship between the field rate of NTSC video when synchronized with the 48kHz sampling rate of audio which results in a whole number of audio samples that can only exist in multiples of five fields.

Fiver: See Senior Spot.

Fix: A change made in a computer program to correct a bug.

Fixed Data Rate Compression: Techniques designed to produce a data stream with a constant data rate.

Fixed Dish System: Satellite system in which the dish does not have to be moved. DirecTV and Dish Network are fixed dish systems.

Fixed Focal Length Lens: See Prime Lens.

Fixed Rate: Information flow at a constant volume over time. See also CBR.

Fixed-Day Spot: A vertically rotating spot scheduled for a specific day of the week.

Fixed-Time Buy: Advertising purchased for a specified, rather than "open" time period, sold at a premium rate.

Fixed-Time-Spot: A spot scheduled for a specific time or specific avail.

Fixing: The removal of unexposed silver halides from the film during processing.

Flag: A large black cloth on a frame used on a shoot to keep light out of part of the composition. In the cutting room it is a small piece of tape attached to a shot in a roll and used exactly as you would use a bookmark. The flag sticks out the side of the roll, making it easy to find that shot again quickly.

Flare: Dark or colored flashes caused by signal overload through extreme light reflections of polished objects or very bright lights.

Flash EPROMs: A non-volatile computer memory chip that can retain information after power is shut off.

Flash Frame: A single frame that is completely clear between two shots. It occurs when the camera is stopped with the gate open, allowing for a very long exposure on that single frame. Rather than a problem, a flash frame can actually be a very helpful thing in the editing room, making it very easy to see where one shot ends and another begins.

Flash Pan: See Swish Pan.

Flashback: A disruption of the chronological presentation of events, in which an event from the past is presented in a program's present.

Flashforward: A scene that breaks the chronological continuity of the main narrative by depicting events which happen in the future.

Flat Lighting: Soft, even lighting that produces minimal shadows and minimizes the depth or dimension in subject matter.

Flat Response: The ability of a microphone or piece of audio gear to reproduce sound equally well over the entire frequency range.

Flat Screen Display: New TV screen technology will reduce the depth of TV sets to that of a framed picture. The TV screens on today's larger sets use bulky cathode ray tubes (CRTs), which are made of a glass envelope and use a controlled beam of electrons striking light-emitting material to display the picture. Flat screen displays are currently in development, and are beginning to show up in some TV sets and computer monitors. For now, CRTs are believed to be the first choice for DTV, but future development will make large DTV and current analog flat-screen displays practical.

Flat: Usually an agreement to perform work or provide a service for a fixed fee or wage which will not be affected by overtime restrictions of unexpected costs. Also used in terms of sets and set construction elements which are generally used to create walls.

Flatbed: An editing machine resembling a desk with a screen in the middle. The film sits flat on plates which are threaded through the center section that has transports for picture and sound.

Flex-Fill: A round cloth bounce card mounted on a flexible ring that can be folded up when not in use.

Flicker: A flicker occurs when the frame rate of the video is too low. It's the same effect produced by an old fluorescent light fixture. The two problems with flicker are that it's distracting and tiring to the eyes. Flicker may also be a problem when static computer images are transferred to video.

Flight: A scheduling tactic which alternates periods of advertising with periods of no activity.

Flip Stand: Device holding camera cards that can be flipped so rapidly that the change is not apparent.

Flip: Special effect in which the picture is either horizontally or vertically reversed.

Float: Periodic vertical movement of the image which occurs as a result of mechanical faults in the camera, printer, or projector.

Floodlight A studio lamp that illuminates a relatively wide area by "flooding" it with light. Also called a Flood.

Floor Manager: Person in charge of the studio during production.

Floor Plan: A pre-production drawing, generally to scale, that indicates all set pieces, walls, and often lights and camera positions.

Flop-over: An optical effect in which the picture is shown reversed from left to right.

Florescent Light Filter: Color correction filters that attempt to give natural color renditions under fluorescent lighting.

Flow: Television's sequence of programs, commercials, news breaks, and so on. The overall flow of television is segmented into small parcels, which often bear little logical connection to one another.

Fluid Head: The pan head of choice for EFP cameras because its internal parts move through a heavy liquid, making very smooth pans and tilts possible.

Fluorescent Lights: You will usually get a greenish tinge off of fluorescent lights. You will want to avoid shooting under them in most situations. However, in recent years many music videos have scenes that were shot under fluorescent lights.

Flutter: Undesirable, rapid change in frequency of an audio or video signal due to variations in tape or disc speed.

Flux: An amount of light which is present as measured in lumens.

Flyaway Units: A portable satellite up-link commonly used in electronic newsgathering.

Flying Erase Head: A moving video erase head, mounted just ahead of the video record head in a video recorder that erases previously recorded video. As opposed to television stationary erase heads, they make possible stable insert edits.

Flying Spot Scanner: See Telecine.

FM (Frequency Modulation): As opposed to AM (amplitude modulation) the form of RF modulation used for television sound transmission. Also used in satellite video transmission and for videotape recording.

Foam Technician: A person responsible for creating foam latex prosthetic appliances from a sculpture created by a makeup artist.

Foamcore: Polystyrene which is sandwiched between paper. It is used for reflectors, soft boxes, and other items because it is stable and easily cut.

Focal Length: Simply put, how wide or narrow a view the lens will provide, smaller numbers being wider and larger numbers being narrower.

Focal Plane: The point behind and perpendicular to a camera lens where a sharp image will appear. This is the target point of a camera CCD where the image is formed.

Focal Point: In a camera lens, that spot where the light rays, bent by the lens, converge before expanding again and striking the film or electronic pickup at the focal plane.

Focus Group: A group of approximately ten to twelve members of the public unrelated to a movie's production who attend a sneak preview. A single focus group is usually composed of a selection of people within the boundaries of a movie's intended audience. The group is extensively questioned by the filmmakers following the screening, and their opinions are incorporated into any further editing that may occur before the premiere.

Focus Pull: The refocusing of a lens during a shot to keep a moving subject in focus or to change the person or object of attention.

Focus: The adjustment of the camera lens so that the image is sharp and clear.

Fog Filter: An optical filter that fits over a camera lens and creates the illusion of fog.

Fog Level: The minimum density of the unexposed area of processed film.

Foldback Speaker: PA-type speaker intended to be heard only by talent. Often used in music concerts so performers can hear accompaniment from surrounding instruments.

Folding Editor: Storyboard mode of displaying clips on the Edit Desk in Quantel editing systems. Each folded clip or segment has its own time-line control.

Foley Artist: A person who creates foley sound effects. Named after an early practitioner.

Foley Mixer: A sound mixer who works with a foley artist to record sound effects.

Foley: A process of sound editing that is completed during the post-production process. Certain sounds, e.g., punching, walking, breaking glass, are added to the final film in this process.

Follow Focus: A shot where focus is changed while shooting to correspond with the moment of the subject (or the camera).

Follow Shot: A moving camera shot which keeps a moving subject within the frame.

Follow Spot: A large, high-powered, stand-mounted spotlight. Typically operated by hand and used to follow action on a stage.

Following Source: The use of lights in a lighting design that is consistent with the apparent sources of light within the room or setting. If a table lamp in a scene is visible and supposedly turned on, subjects near the lamp should be keyed so that the lamp appears to be supplying their illumination.

Foot Candle: Measurement of light. One foot candle is the light of one candle, one foot away. Many light meters will use foot candles as a starting number, which then must be converted into an F-stop based on the sensitivity of the film you are using.

Footage Counters: They are used to show you how much film you have exposed and how much is left to expose. Some footage meters on Panavision cameras have digital readouts accurate to the frame. But most average ones are only approximate to the foot. When you start they are usually set to zero feet. If not you should adjust it to zero.

Footage Encoder Time Code Generator: An electronic device which takes the input from a reader of Keykode numbers, decodes this information and correlates the numbers with the SMPTE time code it generates. These data, along with 3: 2 pull-down status of the transfer, footage count, and audio time code (if applicable) are made available for window burn-ins, VITC-LTC recording and output to a computer.

Footage: Exposed film stock.

Footprint: An area of the earth that is able to receive a particular satellite's signals. This depends on the satellites beam.

For Night: The art of turning day into night, using a sequence of different lenses and filters and special lighting. Distribution. The rights to market and distribute the finished film to different exhibitors for a fee.

Forced Activation Button: Menu buttons that automatically perform the specified action as soon as the button has been highlighted on the menu.

Forced Display: A feature of DVD-Video allowing subpictures to be displayed even if the player's subpicture display mode is turned off. Designed for showing subtitles in a scene where the language is different from the native language of the film.

Forced Selected Button: Menu button that is automatically selected when the menu is displayed.

Foreground: Objects or action which is closest to the camera.

Formalism: An approach to filmmaking or film criticism which emphasizes form over content, arguing that meaning emerges from the way content is presented.

Format Conversion: Process of both encoding/decoding and re-sampling digital rates to change digital data from one format to another.

Format: In film, refers to the film width itself and is measured in millimeters (e.g., super-8, 16mm and 35mm). In videotape, the combination of the width of the tape, measured in inches, (e.g., 1/2," 3/4" and 1") and the process used to store the images on tape (e.g., VHS, Beta). In broadcasting, format refers to the location and timing of breaks and avails in a network programming schedule.

Formula: A familiar plot or pattern of dramatic action which is often repeated or imitated in films, for example, in genres like gangster films and westerns.

FPS: See Frames Per Second.

Fragile Watermark: A watermark designed to be destroyed by any form of copying or encoding other than a bit-for-bit digital copy. Absence of the watermark indicates that a copy has been made.

Fragmentation: Scattering of data over a hard disk caused by successive recording and deletion operations. Generally, this will eventually result in slow data recall, a situation not suitable for video recording or replay.

Frame Accurate: Edit or editing device that is accurate to one video frame.

Frame Animation: Animation procedure in which images are drawn, frame-by-frame and then recorded one-by-one on frames of film or video.

Frame Buffer: Memory used to hold an image for display.

Frame Doubler: A video processor that increases the frame rate (display rate) in order to create a smoother-looking video display.

Frame Enlargement: A photograph of one motion picture frame reproduced from actual footage, not to be confused with a still photographed during the shooting of a scene.

Frame Grabber: A computer expansion board that has the ability to digitize and store a single video frame.

Frame Handles: Extra frames at the beginning and the end of every shot, the exact number will vary from one application to the next, which are used primarily when preparing original material for optical printing, such as the Zero Cut method of blow up, or the creation of a superimposed title, etc. The purpose they serve, in the case of zero cut, is to make sure the registration pin of the printer is not grabbing a splice, which can cause the image to wobble. With opticals they are often used merely to avoid printed-in dirt, which is much more prevalent close to a splice where bits of film cement can flake off.

Frame Line: The small sliver of space between frames.

Frame Rate Conversion: The process of converting video signals from one frame rate to another, usually preserving the number of lines per frame.

Frame Rate: The speed at which the video frames are scanned 30 per second in NTSC video. Movies are created by taking a rapid sequence of pictures (frames) of action and by displaying these frames at the same rate at which they were recorded, the illusion of motion can be created.

The frame rate of a video source is how fast a new still image is available.

Frame Store Synchronizer: Device which can hold a complete video frame in electronic memory. Commonly used to synchronize signals from two or more video sources.

Frame Store: Electronic device that electronically captures and stores complete video pictures.

Frame Synchonizer: A digital electronic device which synchronizes two or more video signals. The frame synchronizer uses one of its inputs as a reference and genlocks the other video signals to the reference's sync and colorburst signals. By delaying the other signals so that each line and field starts at the same time, two or more video images can be blended, wiped and otherwise processed together.

Frame: A single image (of a series of them) on a piece of film. There are 24 frames per second.

Frames Per Second (FPS): The number of still frames (pictures) that give the allusion of motion, which appear in a single second of time.

Framing: Determines what the viewer can and cannot see due to the manipulation of the camera frame (the edge of the image).

Franchise: Contract between cable company and a community specifying a specific level of service in return for a contract for a specific period of time.

Freeze Frame: The continuous repetition of a single frame of video. It is technically not possible for helical videotape recorders to display freeze frames, since they read only one field of video on each pass across the videotape.

French Flag: A small black metal flag attached to the camera with a positionable arm that is used to shade the lens from light in the case of a Flare.

Frequency Discrimination: Exaggeration or diminution of particular frequencies in relation to others.

Frequency Interleaving: The process by which color and brightness signals are combined into one NTSC signal.

Frequency Response: A range of sound frequencies from low to high. A measurement of the limits of microphones, recording and playback machines, and other audio equipment.

Frequency: Sound wave repetition rate in Hz or cycles per second. In electronics, the number of times a signal changes from positive to negative (or vice versa) per second.

Fresnel: A special light-weight lens used in focusing beams of light. Originally used in lighthouses, now also used in high-quality studio and theatrical lights.

Friction Head: Camera mounting head that locks its movement through the action of adjustable friction between moving parts. Sometimes counterbalanced by a strong spring.

Fringe Time: That portion of the broadcast day generally considered to be before or following the television television station's Prime Time and access time (peak viewing time).

Front Lighting: Light source pointed directly at the subject, leaving few shadows.

Front Load: A way to cluster the spots at the beginning of a program, or to the beginning of a broadcast or Standard Week.

Front Porch: This is the area of the video waveform that sits between the start of horizontal blanking and the start of horizontal sync.

Front-screen Projection: As opposed to rear screen projection, an image projected on the camera side of a light-reflecting screen.

F-stop Lens Aperture: The number attained by dividing the focal length of a lens by its aperture. Indicates the amount of light passing through the lens and, subsequently, exposure For example f2.8, f4, f5.6, f8, f11, f16, f22 each have one stop between them, so f2.8 to f16 will be 5 stops.

FTP: See File Transfer Protocol.

Full Focus Lens: An auto-focus lens which can continually (and often automatically) focus from infinity to the macro (extreme close-up) mode.

Full Motion Video: Video that plays at thirty frames per second (NTSC) or 25 frames per second (PAL). G Giga. An SI prefix for denominations of 1 billion (10^9).

Full Screen: A term used to describe the shape of the picture a movie is displayed in order for it to fill a regular (as of 1998) TV screen.

Full Sponsor: The advertiser who buys a program and receives all of the commercial time provided within the program.

Full Value Visual: Television visual which contains a full range of tones from black to white.

Full-Coat: Film which is coated with an iron oxide compound on which sound is recorded and from which sound is reproduced.

Full-track: An audio tape recorder, or recording, that uses the full width of the tape for making recordings.

Fully Scripted Show: A script for a production in which all dialogue is fully written out.

Function: In narrative study, a single action or character attribute. Based in Russian Formalism and the work of Vladimir Propp.

Fundraising Tape: A video designed for use in fundraising presentations, with the clear objective of persuading viewers that this is a cause, campaign or organization worthy of their support. Fundraising tapes can range in style from impressionistic to documentary to direct plea, although viewers have become far too sophisticated for the telethon-style asks of earlier eras.

Funhouse Effect: Objects get wider or narrower or taller or squattier or go through geometric shape distortions as they move across the screen, due to less than perfect "linearity" of circuits and/or "field uniformity" of lenses. The effect is suggestive of images seen in the curved mirrors found at amusement parks, carnivals, or circuses, often in a location called the funhouse.

Fuse: A device designed to interrupt an electrical circuit in the event of an overload of that circuit.

Fuzzy Logic: Computer-based circuitry that is able to use successive approximation to arrive at a result when no clear-cut answer is otherwise attainable.

G

G byte: One billion (10^9) bytes. Not to be confused with GB or gigabyte (2^{30} bytes).

G: A certificate issued by the MPAA indicating that a film is suitable for all ages. See also PG.

GAA: See Gross Average Audience.

Gaffer Grip: A clamp used to attach small lighting instruments to scenery, furniture, doors, and other set pieces.

Gaffer: The crew member principally responsible for transporting, maintaining, and setting up lighting equipment.

Gaffer's Tape: Cloth tape specifically for use on film shoots, usually 2 inches wide in black or silver. The nice thing about gaffer's tape is that, unlike duct tape, it is designed not to leave a sticky residue behind.

Gain Control: A volume-controlling device manipulated by either a sliding fader or a rotating knob.

Gain: Audio or video amplification. It's also the strength of an audio or video signal.

Galaxy Group: The group of companies proposing the Galaxy watermarking format.(IBM/NEC, Hitachi/Pioneer/Sony.)

Gamma Correction: A CCU control which can alter the progression of the black-to-white tonal range. Often used to open up shadow areas in a low-key picture or to hold picture detail in bright video areas.

Gamma: The transfer characteristics of most cameras and displays are nonlinear. For a display, a small change in amplitude when the signal level is small produces a change in the display brightness level, but the same change in amplitude at a high level will not produce the same magnitude of brightness change. This nonlinearity is known as gamma. Adjusting gamma is difficult especially considering that most people cannot tell what is correct unless they have a side by side comparison with the live subject, a supposedly correct photograph of the subject, or a TV that is adjusted correctly. See also Prime Disclaimer.

Gang Synchronizer: A synchronizing device having more than one sprocketted wheel which is used to set up synchronized relationships between film and soundtracks which have been recorded to magnetic film.

Garbage Matte: A matte or key signal used to exclude areas outside the area of interest where it would otherwise be difficult to derive a good key signal, the excluded area being added to the final key signal. The keyer only has to produce a key for the area surrounded by the garbage matte. Typically switcher wipe patterns or lines are used to approximately describe the area of interest. These may be keyframable to follow a moving object.

Gate: The opening on a camera or a projector just behind the lens, through which a single frame is exposed (in the camera) or projected (in the projector).

Gated Microphone: A microphone commonly used in teleconferencing and designed to pick up a predominant sound source (generally, a speaker) while minimizing the input (and ambient noise) from all other microphones.

Gauge: The size, specifically the width, of a film format: 16mm, 35mm, Super-8 are gauges.

GB (Gigabyte): A computer term for one billion units of data. More accurately, 1,073,741,824 bytes.

GBPS(Gigabits/second): Billions (10^9) of bits per second.

GCR: See Ghost Cancellation Reference Signal.

Gear Head: A camera pan head in which gears are used to achieve smooth, controlled pans and tilts. Primarily used in motion picture and single-camera HDTV work.

Gel: A piece of plastic or gelatin placed in front of a light source to change its color.

General Meeting: A "look-see" type of meeting in which a writer meets with a producer, production executive, studio executive, and so forth, as a form of introduction. Generally, in this meeting the producer, production executive, studio executive, and so forth, does not have a specific project in mind for which the writer will be hired.

General Parameters (GPRMs): GPRMs are used to store the users operational history and to modify a players behavior. DVD-Video players have 16 unique GPRMs. Each GRPM may store a fixed length, two-byte numerical value.

General Purpose Interface: See GPI.

Generated Graphics: Graphic material that is generated and/or manipulated by a computer.

Generation Loss: The loss of signal quality incurred in copying material. This can limit the number of generations used in analogue recordings and cause editors to avoid more generations which may make the editing itself more complex, time consuming and limited. Similar rules apply with digital recordings that use compression as many aspects of editing require the signal to be de-compressed and re-compressed: a lossy process. Non-compressed recordings on disks and D1 and D5 tapes incurs essentially no generation losses.

Generation: The number of duplication steps between an original recording and a given copy. A second generation duplicate is a copy of the original master and a third generation duplicate is a copy of a copy of the original master, etc.

Genloc: To reference a signal generator to a signal normally external to and separate from the system controlled by the signal generator. The purpose is to bring a system, such as a studio, into proper timing with an external source, such as a live remote camera, so that special effects such as wipes, keys, and dissolves can be done using the external source in combination with internal sources.

Genre Criticism: A type of film criticism which examines genre films to determine how they reflect or comment on social values.

Genre: A type of motion picture, such as westerns or science-fiction films, which employs similar plots, narrative conventions, character types, and formulas.

Geometric Distortion: The warping or alteration of the true proportions of an image—often just in specific areas.

Geostationary: Satellites orbit the Earth 22,300 miles above the Equator and rotate at the same relative speed and direction as the Earth's surface. Therefore, the satellites appear television stationary. There are nearly 40 satellites currently in this type of orbit over North America, and well over 100 around the globe.

Geosynchronous Satellite: A satellite that orbits at the same speed as the earth and, therefore, stays in the same place in the sky.

German Expressionism: A film movement in Germany from 1919 through the mid-1920s characterized by the use of bizarre decor, lighting, and camera techniques to express strong feelings and inner experiences.

Ghost Cancellation Reference Signal (GCR): A reference signal on (M) NTSC scan lines 19 and 282 and (B, D, G, H, I) PAL scan line 318 that allows the removal of ghosting from TVs. Filtering is employed to process the transmitted GCR signal and determine how to filter the entire video signal to remove the ghosting. ITU-R BT.1124 and ETS 300732 define the standard each country uses. ATSC A/49 also defines the standard for NTSC.

Ghosting: The momentarily overlapping or superimposition of video images generally associated with standards conversion.

GHz (Gigahertz): One billion cycles per second.

GIF (Graphics Interchange Format): Along with the jpg format, a widely-used method of compressing photo and illustration data for Internet use.

Gigabyte: 1 Billion bytes.

Giraffe: A medium-sized microphone boom that extends about 15 feet and is mounted on a rolling tripod base.

Glance-object Cut: A transition from a character looking off-screen to a shot of what the character sees.

Glitch: Any brief electrical disturbance in a picture.

Go Motion: A form of animation similar to stop motion, but which incorporates motion blur. Ordinary stop motion cannot produce motion blur as motion only occurs between frames. Robotic models that are moved during the exposure of each frame produce motion blur, and thus are more realistic. Pioneered by Industrial Light and Magic for Dragonslayer.

Gobo: In video a scenic piece through which the camera shoots and which will become a part of the scene's foreground. For example, by shooting thorough a gobo of a keyhole-shaped piece of black cardboard, the impression is given that the camera is peering through a keyhole. See also Cookie.

GOF: See Group of Audio Frames.

Gofer: Film or video production assistant often sent to "Go for" coffee or other essentials.

GOP: See Group of Pictures.

GPI (General Purpose Interface): This is used for cueing equipment: usually by a contact closure. It is simple, frame accurate and therefore can be easily applied over a wide range of equipment.

Grading: The process of selecting the printing values for color and density for successive scenes in a complete film to produce the desire visual effects.

Graduated Filter: Lens filter that changes its effect from one part of the image to another.

Grain: The silver halide crystals suspended in the emulsion of a piece of film. When struck by light and chemically processed, these crystals change color, resulting in the film image. The smaller the grain, the higher the definition of the image (i.e., the sharper the image).

Graininess: Video noise. An image corrupted by minute, moving particles.

Grand Alliance: U.S. consortium, formed in 1993, to produce "the best of the best" DTV transmission system from among the initially proposed separate systems. Participants were: AT&T, General Instrument Corp., MIT, Philips Consumer Electronics, David Sarnoff Research Center, Thomson Consumer Electronics, and Zenith Electronics Corp. The group eventually proposed the current ATSC digital standard.

Graphic Match: Any juxtaposition of graphically similar images.

Graphics Tablet: Flat computer interface allowing an electronic pen to act like a mouse to control on-screen activities. Unlike a mouse, the electronic pen is pressure sensitive allowing you to vary the amplitude of feedback.

Graphics: All two-dimensional visuals prepared for a television camera, such as title cards, charts, and graphs.

Gray Card: A piece of dull gray cardboard that reflects 18 percent of the light that strikes it. This card is intended as a representation of an object indoors with average reflective potential.

Gray Market: Dealers and distributors who sell equipment without proper authorization from the manufacturer.

Gray Scale Compression: A distortion of successive shades of gray when subject matter is reproduced by video. Typically the brightness values at the darker end of the scale are pushed together so that they merge. This results in most tones, especially skin tones, appearing much darker than they should be.

Gray Scale: A card or test chart containing a range of gray patches from white to black used in setting up gamma or the gray scale response of cameras.

Grazing: A style of television viewing where the viewer switches channels often, either looking for alternative programming or watching multiple programs simultaneously. Also referred to as Channel Surfing.

Green Light: During the process of development of a media program or event, the Green Light point is the point where the funding agent or studio, in most cases, commits to the availability of funds and gives the official "go-ahead" to the production company to commence First Unit photography.

Greensman: A member of the crew who procures, places, and maintains any vegetation on a set.

Grid: A criss-cross arrangement of pipes near the ceiling of a studio used primarily to hang lights.

Grifflon (Griff): A durable material made of three ply high density rubber. The material is attached within a frame and used as either a soft reflector of sunlight or cut or soften direct sun. It generally comes in three sizes: 6' X 6', 12' x 12', and 20' x 20.'

Grip Tape: This is Duct tape style tape, also known as gaffer's tape or cloth tape.

Grip: The crew member principally responsible for the transportation, maintenance and mounting of the camera.

Gross Average Audience (GAA): The estimate which reflects the sum of tuning and viewing minutes to a program. Tuning and viewing to the same minute of program (or its repeat telecast) is counted each time. Typically a GAA is applicable in markets that air repeat telecasts of the same episode.

Gross Rating Points (GRPs): GRPs are the sum of the rating percentage achieved. Since GRPs take into account duplication, the total percentage of all exposures, the number may exceed 100. Expressed as an absolute number, GRPs are referred to as Gross Impressions or Gross Audience.

Ground Row: Row of lights at the bottom of a cyc (generally hidden from cameras) which provide even illumination across the front surface of the cyc.

Group Delay: A phenomenon involving timing differences between video signal components. For example, a long cable run may introduce a substantial delay between the transmission of the color and brightness video information resulting in shadows.

Group of Audio Frames (GOF): The data area of 1/30 second which is composed of 20 audio frames of Linear PCM Audio.

Group of Pictures (GOP): In MPEG video, one or more I pictures followed by P and B pictures. A GOP is the atomic unit of MPEG video access. GOPs are limited in DVD-Video to 18 frames for 525/60 and 15 frames for 625/50.

Guardband: Safety area separating two different segments of a broadcast signal.

Guerrilla Producer: One who produces an effective video on a shoestring budget.

Guillotine: A type of tape splicer which uses imperforated splicing tape.

Gyro Head: A tripod mount containing a gyroscope mechanism designed to minimize camera movement.

Gyrozoom: Gyro stabilized zoom lens used to smooth out or eliminate unwanted camera movement.

H

H.261, H.263: The ITU-T H.261 and H.263 video compression standards were developed to implement video conferencing over ISDN, LANs, regular phone lines, etc. H.261 supports video resolutions of 352 x 288 and 176 x 144 at up to 29.97 frames per second. H.263 supports video resolutions of 1408 x 1152, 704 x 576, 352 x 288, 176 x 144, and 128 x 96 at up to 29.97 frames per second.

HAD: See Hole Accumulated Diode.

Halation: Halation is the effect that occurs when the bright areas of an image appear to softly bleed around the edges of dark areas. This is caused by light going through the emulsion layer, bouncing off the base of the film and exposing the adjacent emulsion. Some film is manufactured with a black anti-halation coating on the base side.

Half Apple: See Apple Box.

Half D1: An MPEG-2 video encoding mode in which half the horizontal resolution is sampled (352x480 for NTSC, 352x576 for PAL).

Half-track: An audio tape recorder or recording that uses one-half the width of the tape for recording an audio signal. The tape can then be reversed and the other half on the tape used during a reverse pass.

Hall of Mirrors (HOM): A description of an effect some VIS errors cause in the level, where a portion of a wall (ceiling, floor, etc) is not drawn properly.

Halogen: This is the gas contained in the lamp of a Quartz Light, which prolongs the life of the tungsten filament. Quartz Lights are sometimes called Halogen Lights for this reason.

Hammock Position: Programming strategy in which a weak show is placed between two stronger shows in an attempt to boost ratings.

Hand Props: Any object on a set which is used or handled by the talent. Also called a Property.

Hand Signals: Production information communicated silently by the hands to on-camera talent by a floor director.

Handheld: Shooting without a tripod, but with the camera held by the cameraperson.

Handle: An extra number of frames attached to the head and tail of an optical print as a safety precaution.

Hanging Dots: Cross-luminance near sharp saturated horizontal colored edges.

Hard Coded: Data in a computer program that can be changed only by a programmer.

Hard Copy: As opposed to soft copy, a copy of text or data printed out on paper.

Hard Disc Recording: Digitally recording audio or video on a computer-type disk drive system.

Hard Light: Hard light is usually from a key light and as the main source of illumination for most scenes it casts bold shadows and gives you the impression that the light in a scene comes from one particular direction.

Hard Matte: Refers to widescreen movies that are filmed using a camera that blocks off the top and bottom edges of the 4:3 aspect ratio film frame, recording an image that has the aspect ratio of the finished picture and leaving unused the top and bottom edges of each frame on the film. (This is how panoramic 35mm still cameras work.) Some producers use this technique so that the theater projectionist cannot err (or deliberately modify the presentation) by opening up the projector's aperture plates too much.

Hard News: Refers to news stories that examine events which affect society as a whole (e.g., national politics and international relations).

Hard Wired: An electronic piece of equipment that is more or less permanently wired to other equipment.

Hard-copy Prompter: A camera prompting device that incorporates a moving paper script. See also Soft-copy Prompter.

Hardtop: Slang for a normal indoor theatre. See also Ozoner.

Hardware: The physical components of a computer and its associated equipment.

Harmonic Distortion: Acoustic distortion characterized by unwanted changes between input and output at a given frequency.

Harmonics: Acoustical or electrical frequencies that are multiples of the basic frequency or musical tone.

HAVi: A consumer electronics industry standard for interoperability between digital audio and video devices connected via a network in the consumer's home.

Hays Production Code: In the 1920s, the American public became alarmed at the increasingly frequent portrayal of violence, sex, and lawlessness on movie screens. Wishing to avoid government regulation, the Motion Picture Producers and Distributors of America created their own regulatory body and appointed Postmaster General Will H. Hays as head. His influence became so great that this body became known as the "Hays Office." The Hays Production Code for Motion Pictures was introduced in 1934, and by today's standards was extremely strict. It was mainly concerned with violence and sex, but had references to crime in general. Hays Code was abolished in 1967. See also Blacklisting.

Haze Filter: A filter that removes ultraviolet light and diminishes haze in a scene.

Hazeltine: A machine which a color time at a lab or optical house uses to determine how to "time" a film print for the proper amounts of red, blue, and green light. Also called a color analyzer.

HD-CIF: See CIF.

HD-SDTI: High Data-Rate Serial Data Transport Interface, defined by SMPTE 348M.

HDTV: See High Definition Television.

Head Clog: A loss of video signal resulting from dirt in the microscopic gap of a video head.

Head Drum: Head end Initial portion of cable television distribution system where signals are collected and initially amplified.

Head Room: The space between the top of a subject's head and the top of the video frame.

Head: The beginning of a shot or a roll is called the head. The uppermost portion of a tripod or pedestal which provides for the ability to pan and tilt the camera.

Head-cleaning Cassette: A cassette loaded with material which cleans the video heads as the cassette is "played." Often the cleaning material is augmented with a cleaning fluid.

Headend: A cable system's control center where incoming signals from satellite and other sources are converted and put on outgoing channels to subscriber's homes.

Headphones: Listening device that fits over the head and covers the ears used to listen to audio sources.

Helical Scan: A method of recording video information diagonally on a tape, used in home and professional VCRs. High speed rotating video heads scan these diagonal video tracks, giving an effective tape speed much higher than the actual tape speed allowing more information to be recorded on a given length of magnetic tape.

Hertz: See Hz.

Hexadecimal: A numbering system whereby place values are based on powers of 16 (base 16). It uses the digits 0-9 and the letters A-F to represent values 0 through 15 in each position. For example: decimal 10 = hexadecimal A; decimal 30 = hexadecimal 1E. The hexadecimal numbering system is used to abbreviate binary data. See also Binary.

Hiatus: The interruption in an advertiser's media schedule between flights. Sometimes, the temporary removing of a television show from the programming schedule.

HFS (Hierarchical File System): A file system used by Apple Computer's Mac OS operating system.

Hi Hat: This is a square of plywood with a bracket attached, to which a tripod head may be added (or is sometimes permanently affixed) used for filming with the camera very low to the ground. Its name is a bit of a contradiction, to its use nowadays, but it used to be that a Hi Hat was for shooting from very high up, with the plywood board being mounted up high somewhere.

HI-8: An improved version of the 8mm tape format capable of recording better picture resolution (definition). A higher-density tape is required which provides a wider luminance bandwidth, resulting in sharper picture quality (over 400 horizontal lines vs. 240 for standard 8mm) and improved signal-to-noise ratio. Camcorders using this format are very small, light and provide a picture quality similar to S-VHS.

Hi-Con: Generally a high contrast film print which provides the maximum contrast between the light and dark elements.

Hierarchical File System: See HFS.

HI-FI (High Fidelity): Most commonly used to refer to the high quality audio tracks recorded by many VCRs. These tracks provide audio quality approaching that of a CD. However, because they are combined with the video signal before recording, audio dubs using them are impossible without re-recording the video.

High Angle: A shot in which the camera is placed higher than the filmed actor or object, so that the camera looks down on the actor or object.

High Concept: Phrase connected with scripts which have a premise or storyline which is easily reduced to a simple and appealing one line.

High Definition Television (HDTV): Television capable of displaying at least 720 progressive or 1080 interlaced active scan lines. It must be capable of displaying a 16**:** 9 image using at least 540 progressive or 810 interlaced active scan lines.

High Density: Videotape or computer disks that pack more magnetic particles per square inch and thus allow a greater concentration of data.

Highboy: A heavy-duty rolling stand, usually with a combo head, that has a junior receiver and a large grip head. Also called Overhead Stands.

High-contrast Lighting: A style of film lighting which creates a stark contrast between bright light and heavy shadows. See also high-key lighting and low-key lighting. high-key lighting A style of film lighting which creates bright, even illumination and few conspicuous shadows. See also high-contrast lighting and low-key lighting.

High-key Lighting: A style of lighting where the ratio of key light to fill light is small, creating bright, even illumination, with few shadows and a minimum of contrast.

Highlight Information (HLI): This is used to specify button highlights for menus. HLI contains information on the button number, highlight timing, palette for sub-picture highlights, coordinates of the button, etc.

Highlights: The brightest parts of a video image.

High-Pass Filter: An electronic filter used in various audio circuits to attenuate all frequencies below a chosen frequency.

High-speed Film: A rough-grain stock that registers a limited range of gradations between black and white. Requires the least amount of light.

Hip-Pocketing: Agent/agency's practice in which an individual(s) is represented by the agent/agency on a single project only, with no agreement that the agency or agent will continue to represent the individual once the project or interest in the project has ended.

Hiss: The background noise generated in an audio system which is internally generated by microphones, amplifiers, and tape.

Historical World (Historical Reality): The reality that is processed, selected, ordered, and interpreted by nonfiction television programs.

HLI: See Highlight Information.

HMI Light: A high-efficiency, discharge-type light source that has the same color temperature as sunlight.

HMI: Halogen Metal Incandescence. See HMI Light.

Hold: A word used on a continuity report to indicate that a particular take should be kept, but not developed.

Hole-Accumulated Diode (HAD): A CCD sensor structure designed to suppress certain types of noise inherent to CCDs. See also CCD.

Hollywood Box: A stage plug-type box without fuses.

Hollywood Montage: See American Montage.

Hollywood Realism: See Realism.

HOM: See Hall of Mirrors.

Homage: Usually a respectful tribute to someone; in the movies this can often occur when within one movie reference is made to another.

Hook: An attention-getting incident in a production designed to capture an audience's attention.

Horizontal Blanking: The period in the scanning of the TV picture when the electron beam is momentarily extinguished as it moves from right to left. During this period the beam can return unseen to scan another line.

Horizontal Resolution: The ability of horizontal scanning lines of a video system to reproduce picture detail. This is based primarily on how quickly the system can appropriately respond to changes in image intensity. In broadcasting, the movement of a spot within a specified day-of-week range on subsequent airings of the spot schedule. The computer assigns a weekday for each spot to ensure an even distribution of different days throughout the schedule. For example, a spot scheduled at 10:30 am Monday through Friday could move from 10:30 am Monday to 10:30 am on Wednesday to 10:30 am on Tuesday to 10:30 am on Friday to 10:30 am on Thursday in five successive weeks.

Horizontal Retrace Interval: The time during which the electron beam is turned almost off (as if to draw black) and moved from right to left to get ready to draw the next scan line on the picture tube. In NTSC broadcast video, 10.9 microseconds are set aside for this purpose; the entire line, retrace interval and all, takes 63.5 microseconds. The time it takes to draw an entire line is adequate to allow the electron beam to draw at most 530 dots alternating black and white, given a 4.2 MHz bandwidth; the horizontal retrace interval takes away about 90 of these dots. Of the remaining 440 (give or take a few), 330 fit in the largest circle that in turn fits in the 4:3 screen, which leads to the 330 line horizontal resolution that is often mentioned when it comes to broadcast TV. The exact number of dots used for the picture material depends on the program source and the make and model of equipment.

Horizontal Scan Rate: This is how fast the scanning beam in a display sweeps from side to side. In the NTSC system, this rate is 63.556 ms, or 15.734 kHz. That means the scanning beam moves from side to side 15,734 times a second.

Horizontal Sync: That portion of the sync signal that controls the horizontal timing (and therefore horizontal location) of each line of picture. See also Blanking Level.

Horse: A support for one or more rolls of film used on a cutting table.

Hot Set: A set on which a scene is in the process of being shot; labeled thus to indicate that it should not be changed or disturbed.

Hot Splicer: A film splicing machine of precision construction in which portions of a film are overlapped, cemented, then warmed and dried by a heating unit.

Hot Spot: Undesirable concentrations of light on a set or subject matter.

Hot: Attributed characteristic of a wire carrying an electrical current. The term is also used to signify instrument that is turned on for example, a "hot camera" or "a hot microphone."

House Lights: Normal, existing overhead work lights in the studio or theater.

House Sync: An internal timing reference used to synchronize all transports within a facility.

Households Using Television (HUT): HUT Level is the percentage of all television households tuned to any programming during a specific time of day. While VCR recording contributes to a HUT Level (and a program rating), VCR playback and playing video games are not included in the HUT.

Households: Group of individuals occupying a house, apartment, group of rooms or a single room that is considered a housing unit.

HSI: HSI stands for Hue, Saturation and Intensity.

HSL: This is similar to HSI, except that HSL stands for Hue, Saturation, and Lightness.

HSV: This is similar to HSI, except that HSV stands for Hue, Saturation, and Value.

HSYNC: See Horizontal Sync.

Hue: A specific color from within the visible spectrum of white light: e.g., red, green, blue.

Huffman Coding: A lossless compression technique of assigning variable-length codes to a known set of values. Values occurring most frequently are assigned the shortest codes. MPEG uses a variation of Huffman coding with fixed code tables, often called variable-length coding (VLC).

Hum Bar: Visible bar in video caused by introduction of 60-Hz electrical interference.

Hum: Unwanted low frequency audio noise caused by improperly shielded or improperly grounded audio cables and circuits.

Human Interest: In contrast to hard news, stories in which news value is secondary to an appeal to feelings, emotions or sensations.

HUT: See Households Using Television.

Hybrid Editing: The marriage of disk and tape within an edit suite. There are two different approaches: adding disks to a linear suite or incorporating tape into a disk editing system. The first does not deliver any of the benefits of nonlinear editing and constrains the disks to working in a linear fashion. Quantel edit systems take the second approach and allow the straightforward inclusion of material from tape within the nonlinear edit.

Hype: Overzealous praise or advertising.

Hyperband: CATV channels above the basic channels.

Hypercardioid Microphone: A highly unidirectional microphone, for which the pickup pattern is narrower than that of a cardioid microphone. So-called "shotgun" microphones have a hypercardioid pattern.

Hyperfocal Distance: The hyperfocal distance is a distance set on the focusing ring of the lens that will most efficiently use the Depth of Field present. A depth of field chart will list possible distances and graph out the area of focus at different F-stops. There does not necessarily have to be a subject to focus on at that distance.

Hypodermic Needle Theory: An effects theory which purports that the viewer is passive and directly and immediately affected by what he/she sees on television.

HZ: Cycles per second; a measure of electromagnetic frequency representing the number of complete electrical waves in a second. One kilohertz (kHz) is one thousand cycles per second.

I

I Picture: In MPEG video, an intra picture that is encoded independent from other pictures (see intraframe). Transform coding (DCT, quantization, and VLC) is used with no motion compensation, resulting in only moderate compression. I pictures provide a reference point for dependent P pictures and B pictures and allow random access into the compressed video stream. Also called I Frame.

I Signal: One of the two color signals, containing reddish orange and bluish green components to which the human eye is sensitive.

i.Link Trademarked Sony name for IEEE 1394.

I/O Board: An electronic board that can transfer information from a computer to related devices.

I/O: "Input/output." Typically refers to sending data to and from devices.

IATSE: See International Alliance of Theatrical and Stage Employees.

IBEW: See International Brotherhood of Electrical Workers.

IC: See Integrated Circuit.

Icon: Symbolic images in software displays that users can select to carry out desired tasks. Generally speaking, an icon is an object that represents a theme or an aspect of the character or the like. In the specific context of semiotics, a type of sign, wherein the signifier physically resembles the signified. For example, a photograph (signifier) is a mechanical reproduction of what is photographed (signified).

Iconography: The use of a well-known symbol or icon. For example, in the movies a star's persona can be iconographic. Iconography can also involve specific stylistic conventions of a particular genre or the use of archetypes or stylistic lighting, costuming, etc.

ID Error Correction (IEC): 2 special error correction bytes that are added to each sector header.

ID: See Identification Data.

IDE (Integrated Drive Electronics): An internal bus, or standard electronic interface between a computer and internal block storage devices. IDE was adopted as a standard by ANSI in November 1990. ANSI calls it Advanced Technology Attachment (ATA). Also see E-IDE and ATAPI.

Idealism: Treatment of subject matter in a work of art (play, novel, film, etc.) in which the concept of beauty and/or form is stressed, characterized usually by the selection of actors and/or characters which meet a specified standard of perfection.

Identification Data (ID): 32-bit field identifying the sector number within the disc volume.

Ideological Criticism: An area of television criticism, concerned with class and gender representation, that studies society's competing discourses and the position of the individual within society.

IDTV: See Improved Definition Television.

IEC 60461: Defines the longitudinal (LTC) and vertical interval (VITC) time code for NTSC.

IEC 60958: Defines a serial digital audio interface for consumer (SPDIF) and professional applications.

IEC 61834: Defines the DV (originally the "Blue Book") standard. Purchase the specification. Also see SMPTE 314M.

IEC 61880: Defines the widescreen signaling (WSS) information for NTSC video signals. WSS may be present on lines 20 and 283.

IEC 61883: Defines the methods for transferring data, audio, DV and MPEG-2 data over IEEE 1394.

IEC 62107: Defines the Super VideoCD standard.

IEC: International Electrotechnical Commission.

IEC: See ID Error Correction.

IED: ID error correction. An error-detection code applied to each sector ID on a DVD disc.

IEEE 1394: A high-speed "daisy-chained" serial interface. Digital audio, video, and data can be transferred with either a guaranteed bandwidth or a guaranteed latency. It is hot-pluggable, and uses a small 6-pin or 4-pin connector, with the 6-pin connector providing power.

IEEE: See Institute of Electrical and Electronics Engineers.

IFB: See Interrupted Feedback System.

IFE: In-flight entertainment.

Illegal Video: Some colors that exist in the RGB color space can't be represented in the NTSC and PAL video domain. For example, 100% saturated red in the RGB space (which is the red color on full strength and the blue and green colors turned off) can't exist in the NTSC video signal, due to color bandwidth limitations. The NTSC encoder must be able to determine that an illegal color is being generated and stop that from occurring, since it may cause over-saturation and blooming.

Illusion of Depth: The ability of the two-dimensional television image to create an illusion whereby space seems to recede into the image. A telephoto lens creates a small illusion of depth and a wide-angle lens creates a large one.

Image Buffer: For all practical purposes, an image buffer is the same as a frame buffer. An image is acquired and stored in the image buffer. Once it is in the image buffer, it can typically be annotated with text or graphics or manipulated in some way, just like anything else in a frame buffer.

Image Capture Board: Computer circuit that stores and displays images for computer manipulation.

Image Compression: Image compression is used to reduce the amount of memory required to store an image. For example, an image that has a resolution of 640 x 480 and is in the RGB color space at 8 bits per color, requiring 900 KB of storage. If this image can be compressed at a compression ratio of 20: 1, then the amount of storage required is only 45 KB. There are several methods of image compression, but the most popular are JPEG and MPEG. H.261 and H.263 are the video compression standards used for video conferencing.

Image Enhancer: An electronic circuit that sharpens detail in video pictures. Image enhancers commonly improve edge sharpness, reduce color edge fringing, correct color displacement, and reduce snow and noise.

Image Intensifier: An electronic attachment to a lens system that amplifies incoming light. Also called Night Viewing Device.

Image Stabilizer: A device that is part of a camera lens or camera mount that reduces or cancels vibration or moderate movement. Often used when a camera is hand-held or mounted in a moving vehicle.

Impedance Matching: A video signal occupies a wide spectrum of frequencies, from nearly DC (0 Hz) to 6 MHz. If the output impedance of either the video source, cable or input impedance of the receiving equipment are not properly matched, a series of problems may arise. Loss of high frequency detail and color information as well as image instability, oscillations, snow, ghost images and component heat-up may result. Proper connections and cable types provide correct impedances. See also Load Resistance.

Impedance: A unit of electrical resistance. Generally expressed as high impedance (hi-Z) or low impedance (low-Z). The impedance of various pieces of interconnected video and audio equipment must match or else various types of distortion and inefficiency will result.

I-MPEG (Intraframe MPEG): An unofficial variation of MPEG video encoding that uses only intraframe compression. I-MPEG is used by DV equipment.

Impression: The delivery of one advertisement to one person. Derived from Ratings, impressions are a statistical figure based on the estimated number of people in a demographic area.

Improved Definition Television: IDTV is different from HDTV. IDTV is a system that improves the display on TVs by adding processing in the TV; standard NTSC or PAL signals are transmitted.

Improvisation: Technique of method acting style used mostly in rehearsal; the actor puts him/herself into the mind of the character, places the character into imagined situations and proceeds to invent dialogue and action.

Impulse Noise: Artifact of video signal transmission. It may be suppressed by median filtering. "Sparklies" is the term for the specific form of impulse noise typical of marginal satellite reception.

Impulse Pay Per View: The ability to buy a particular program on a last-minute decision. Pushing the "buy" button on the remote control instead of having to make a phone call.

In The Can: Same as "that's a wrap" to indicate that the scene or program which has been completed. Historically, it would refer to the act of placing exposed motion picture film back in its can so that it can be sent to the laboratory for developing.

In-camera Editing: Editing done within the camera itself by selectively starting and stopping the camera for each shot. Also called Editing in the Camera.

Incident Light Reading: An incident light reading measures the amount of light hitting the subject. You take an incident reading with a light meter equipped with a white half-sphere which acts as a stand-in for the subject. The sphere is pointed at the camera, so that the same light hitting the subject is hitting the sphere. The other type of light reading is a Reflective Light Reading.

Incident Light: Light striking a subject, as opposed to light reflected by a subject.

Incident Meter: As opposed to a reflected light meter, a photoelectric meter that measures and intensity of light falling on an object.

Incoming Scene: The second scene to appear in a dissolve or wipe effect.

Independent Film: A movie not produced by a major studio. Also called an Indie.

Independent Producer: A producer who is not affiliated with a major studio.

Independent Television Station: A broadcast television station which is not affiliated with a network.

Index: A log of reference points within an audio or video piece for the purpose of searching the media.

Indexing: A process that logs reference points within an audio or video file for the purpose of searching the media.

Indie: A movie not produced by a major studio.

Indigenous Sound: Sound or music originating from a source apparent within a film scene. See Foreground Music.

Inductance: The resistance of a coil of wire to rapidly fluctuating currents which increases with frequency.

Infinity: The furthest distance on the focusing ring of a lens.

Infomercial: A commercial typically 90 seconds or more in length designed to supply information about a product or service as opposed to a specific sales message.

Infrared: Energy near visible light (beyond visible red) used by some types of film and by some autofocus camera devices.

Ingénue: A naive girl or young woman in a dramatic story.

Ink: To sign a contract.

Inkie: A small (250 watt) fresnel type light.

In-line Amplifier: Amplifier inserted between two wires that increases the amplitude of a signal. input selector switch A multi-position switch found on many audio control boards that connects any one of several sources to a specific audio channel.

Inning: A period of time.

In-point: The point from which the video and/or audio is to be used in an edit.

Input: Data to be processed, or the act of keying in data.

Insert Edit: An electronic edit in which the existing control track is not replaced during the editing process.

Insert Editing (tape): A mode of editing on tape by recording, or dropping-in, new video and audio over existing material. This is the most common mode of videotape editing and is the alternative to assemble editing. The existing control track and time code already recorded on the tape remain undisturbed as only the new video and audio are inserted, minimizing the chance of any discontinuity in the edited result. Insert editing is largely irrelevant in nonlinear editing systems where the term has taken on a new and, perhaps, more intuitive meaning. See also Assemble Editing.

Insert Shot: A close-up of some detail in the scene. (Sort of like a cutaway without the "-away" aspect.)

Insert: A shot or sequence inserted into a existing edit package.

Insertion Loss: Loss of signal strength due to connection to a device.

Instant Replay: Immediately repeating a segment of a televised event for viewers to see again—often in slow motion.

Institute of Electrical and Electronics Engineers (IEEE): An organization responsible for creating a number of technical standards in television.

Institutional Video: Television production not intended for a mass audience or for general broadcast. Includes most corporate, educational and medical productions.

Instructed Spots: Logged spots that have copy instructions.

Integrated Circuit (IC): A very small electronic component consisting of scores of transistors, resistors, capacitors and diodes all in one sealed housing.

Integrated Drive Electronics: See IDE.

Integrated Receiver Decoder (IRD): A satellite receiver with a built-in decoder for unscrambling subscription channels. It is usually called the receiver.

Integrated Services Digital Network (ISDN): Allows computers to communicate over existing phone lines using a digital telephone network at much higher speeds than are possible with an analog modem.

Integrated Services Digital Network: See ISDN.

Intellectual Montage: Editing intended to convey an abstract or intellectual concept by juxtaposing concrete images which suggest it.

Intensity: The brightness of a light source as measured in foot-candles or lux.

Interactive Cable: Cable TV capable of sending information and responding to input from a user.

Interactive Mode: Type of television text in which the historical world is mixed with that of the video/film maker—according to Bill Nichols's approach to nonfiction television and film. This occurs in one of two ways. The social actor is brought into a television studio; and/or a representative of television enters the historical world to provoke a response from social actors. In another context, interactive is coming to refer to the capacity of the viewer to respond to or affect what is seen on television, for example, through home shopping services.

Interactive TV: An interactive television service that lets you use the enclosed remote control to access up-to-the-minute news, sports, financial information, weather, get program trivia, respond to free offers and shop, all while you watch TV.

Interactive Video: The video presentation in which the user's actions, choices and decisions affect the course of the presentation.

Intercast: A method developed by Intel for transmitting web pages during the vertical blanking interval of a NTSC or PAL video signal. It is based on NABTS for (M) NTSC systems.

Intercom: See PL Line.

Intercutting: An editing method whereby related shots are inserted into a series of other shots for the purpose of contrast or for some other effect.

Interface: The connection between two devices, such as a computer and a peripheral.

Interframe: Something that occurs between multiple frames of video. Interframe compression takes temporal redundancy into account.

Interior Monologue: See Monologue.

Interior Scenes: Scenes set inside, in particular on studio sets, though also including location interiors.

Interlace: A process in which the picture is split into two fields by sending all the odd numbered lines to field one and all the even numbered lines to field two. This was necessary in the early days of TV when there was not enough bandwidth to send a complete frame fast enough to create a non-flickering image.

Interleave: To arrange data in alternating chunks so that selected parts can be extracted while other parts are skipped over, or so that each chunk carries a piece of a different data stream. In DVD, used for seamless multi-angle and Director's cut features, in which multiplexed streams are subsequently interleaved to allow seamless playback of alternate program material.

Interlock Projector: A projector used to reproduce the picture while synchronized sound is played back on an accompanying machine or other linked sound device.

Interlocked: Two or more devices (most commonly dubbers in a mixing facility) with motors that run in sync are interlocked. It is not quite correct to say that a sync sound camera and tape recorder are interlocked, regardless of whether they use crystal of cable sync, since the tape recorder is recording pilottone and not really running with its motor interlocked with the camera motor.

Intermediates: A general term for color masters or duplicates.

Intermodulation Distortion: An amplitude change in which sum and difference tones (harmonics) are present in the recorded signal.

Internal Key: As opposed to external key, with internal key the cutout portion of a background video source is filled with a signal from the switcher.

Internal Sync: Synchronizing signals generated by a camera, recorder, or other picture source without reference to or need of external synchronizing signals.

Internal Telecommunications Union: See ICU.

International Alliance of Theatrical and Stage Employees (IATSE): A trade union for various behind-the-scenes workers.

International Brotherhood of Electrical Workers (IBEW): A trade union of electrical workers.

International Radio Consultative Committee (CCIR): An international standards committee no longer in operation and replaced by the International Telecommunications Union (ITU).

International Telecommunications Union (ITU): An international broadcast standards committee that replaced the CCIR.

Internegative: An intermediate copy of a film, made on a very fine-grained stock, and used to make a greater number of prints than it is practical to make from the A&B Rolls.

Interpolation: In digital video, the creation of new pixels in the image by some method of mathematically manipulating the values of neighboring pixels. Interpolation involves using existing sampled data points to predict values between these points. The simplest form of interpolation is to connect these data points with straight lines (a two-tap linear interpolation filter). It is used in digital video system for picture size changes and other manipulations including standards conversion.

Interpositive: An intermediate copy of a film, made on a very fine-grained stock, usually required as an intermediate step to making an internegative.

Interrupted Feedback System (IFB): A small earpiece worn by talent that carries program sound or (when interrupted) instructions from a director.

Intertextuality: The intertextual, self-reflexive quality—as when one television text (e.g., a commercial) refers to another (e.g., a program or commercial) or to other types of media texts.

Intertitles: Title card appearing intercut with a scene.

Intervalometer: A device that attaches to the camera for filming single exposures, much like an animation motor, except that an intervalometer is capable of exposing single frames automatically, as in the technique of Time Lapse photography.

Interview Format: Using a basic interview as the basis of production or program segment.

Intensity: This is the same thing as brightness.

Intraframe MPEG: See I-MPEG.

Intraframe: Something that occurs within a single frame of video. Intraframe compression does not reduce temporal redundancy, but allows each frame to be independently manipulated or accessed.

Intrinsic Interest: An area of the film image which demands the viewers attention because of its dramatic importance but which is not the visual centre of the image.

Introduction: The beginning or introductory portion of a production or production segment.

Intrusion on Seclusion: Related to invasion of privacy, a legal term denoting the unlawful entering into a place or state without invitation, right or welcome. It may actual, such as trespassing on private property to secure a news story, or figurative, such as when a telephoto lens or electronic eavesdropping device is used.

Invar: Trademark for a metal alloy used for picture tube shadow masks, which expands and contracts due to temperature changes less than most metals. Used so that more brightness can be obtained which requires stronger electron beams which in turn cause the shadow mask to heat up more and where expansion causes the holes to go out of alignment with the phosphor dots or stripes causing discoloration.

Inverse Telecine: A process that removes interlaced video fields added in the Telecine process to view film footage in media file format. The result will be video at 24 fps without interpolated fields. This process is necessary to properly prepare Telecined footage for playback on progressive scan devices, such as computer screens.

Invisible Editing: Editing made unobtrusive by carefully cutting on action or matching action between shots. Also called Invisible Cutting.

IP/IN: The process of making a new negative of a film by striking an interpositive and then, from that, striking a new negative (internegative).

IPS (Inches Per Second): Generally refers to tape speed or tape writing speed.

I-Q: Refers to one method of encoding of the color content of a video signal such that all colors could be depicted using a two dimensional diagram, more specifically a color wheel. There are two color signals one (I) of which represents oranges and blues and the other (Q) represents greens and purples. (Both together are used to represent other colors.)

IRD: See Integrated Receiver Decoder.

IRE Scale: The scale to determine video signal amplitudes devised by the Institute of Radio Engineers (IRE), an American organization now called the Institute of Electrical and Electronic Engineers (IEEE). The IRE scale includes a total of 140 units, with 100 up and 40 down from zero.

IRE Unit: An arbitrary unit used to describe the amplitude characteristics of a video signal. White is defined to be 100 IRE and the blanking level is defined to be 0 IRE.

Iris Wipe: A wipe effect in the form of an expanding or diminishing circle.

Iris: Like the iris of the eye, a valve within a lens to control the amount of light that passes through. Opening the iris permits more light to pass through the lens and closing the iris less. The degree to which the iris is open or closed is measured in F-stops, and on some lenses supplemented by T-Stops.

Iris-in: A shot that opens from darkness in an expanding circle of light.

Iris-out: The opposite of an iris-in.

iScan: Trademark of Silicon Image Corp. for its line of stand alone de-interlacers, also usable as NTSC to VGA converters.

ISO 2202: Information Processing—ISO 7-bit and 8-bit coded character sets.

ISO 3166: Codes for the representation of names of countries.

ISO 3901: Documentation—International Standard Recording Code (ISRC).

ISO 639: Codes for the representation of names of languages.

ISO 8859-1: Information Processing—8-bit single-byte coded graphic character sets.

ISO 9660: The international standard for the file system used by CD-ROM. Allows file names of only 8 characters plus a 3-character extension.

ISO/IEC 11172: Information technology coding of moving pictures and associated audio for digital storage media up to about 1.5 Mbit/s. (MPEG-1)

ISO/IEC 13818: Information technology—generic coding of moving pictures and associated audio. (MPEG-2)

ISO/IEC DIS 13818-3: Information technology—generic coding of moving pictures and associated audio.

ISO: International Standards Organization.

Isolated Camera: A camera that is both routed through the switcher and has its own separate, VTR. The recorded signal from an isod camera is typically used for instant replays.

ISP (Internet Service Provider): A company that provides commercial and consumer access to the internet through dial-up and leased line access.

ISRC: International Standard Recording Code.

Italian Neo-realism See Neo-realism.

ITS: The International Teleproduction Society.

It's a Wrap!: What to say when you are done shooting,

either for the day, at that particular set, or on the entire film. Usually if it's not the final shoot you would say you are just going to "wrap for the day."

ITU-R BT.601.5: Formerly known as CCIR 601. An international standard for component digital television from which was derived the SMPTE 125M standard. This ITU recommendation defines the sampling systems, matrix values and filter characteristics for both Y, B-Y, R-Y and RCB component digital television.

ITU-R BT.601: The international standard specifying the format of digital component video. Currently at version 5 (identified as 601-5).

ITVA: The International Television Association.

J

Jack: A female electrical connection typically a phone-plug receptacle. Also, a brace for scenery.

Jaggies: Jargon for spatial aliasing that appears as jagged steps on diagonal lines in a TV picture. Caused by insufficient filtering, violation of the Nyquist limit and/or poor interpolation. See also Aliasing.

Jam Sync: A process of locking a time code generator to an existing coded tape in order to extend or replace the code, used when code is of poor quality.

Japanese Lantern: A paper-covered wire frame globe into which a socket and bulb may be placed.

JCIC: Joint Committee on Intersociety Coordination.

JEC: Joint Engineering Committee of EIA and NCTA.

Jib Arm: A mechanical are which is supported on a dolly, tripod, or other device, which is counterweighted to hold a camera for an increased range of motion.

JIP: See Joined In Progress.

Jitter: Small, rapid variations in a waveform or image due most often to mechanical disturbances.

J-Lar: A transparent tape which is used to splice gels together.

Jog: The slow, frame-by-frame viewing of a videotape, generally to make edit decisions.

Joined In Progress (JIP): The act of partially airing a program due to a run-over of the prior program or cut-in news segment.

Joystick: A hand-operated control which can be moved in any of four directions (sometimes even vertically) to control various computer operations.

JPEG (Joint Photographic Experts Group): JPEG is a digital compression standard for still video images that allows the image to occupy less memory or disk space. Like the MPEG standard, it includes options for trading off between storage space and image quality.

Judder: An instability introduced when images sampled at one frame rate are converted to a different frame rate for viewing. This effect is most noticeable when frames are repeated or deleted in order to obtain slow motion or fast motion. See also Motion Artifact.

Juicer: An electrician.

Jump Cut: A cut edit where the out and in footage are both from the same shot or the action of the outgoing shot does not match that of the incoming. Any changes between the two will show as a jump: hence the name. This can be used as a production style or can be softened using mixes (easily done in a disk-based suite) but in both cases the edit is obvious. A common device to cover the discontinuity is the use of cut-aways.

Junior: A 2K fresnel light unit. It may also mean any 1 1/8 inch spud or mounting pin or any 1 1/8 inch female receiver.

Juxtaposition: Act of putting together things close together or side by side.

K

K byte: One thousand (10^3) bytes. Not to be confused with KB or kilobyte (2^{10} bytes). Note the small "k."

K: An abbreviation for Kilowatts. There are 1,000 Watts in 1 Kilowatt. It is used when talking about quartz lights or HMIs, as a way to measure their brightness based on their power consumption. A "1K" is a 1,000 Watt light, a "2K" a 2,000 Watt light, etc. "K" is also an abbreviation for Kelvin, such as 3,200K for tungsten balance, 5,400K for daylight, etc.

KB (Kilobyte): A computer term for a unit of data. Although commonly translated as 1,000 bytes, it is actually 1,024 bytes.

Kbps (Kilobits/second): Thousands (10^3) of bits per second.

Kell Factor: The ratio of the number of psychologically perceived analog lines of resolution to the number of scan lines in interlaced video compared with the perceived resolution of the same picture reproduced with progressively scanned video. The larger the better, 0.7 is considered very good. Deficiencies such as scan lines too thin or too thick, and interlace flickering due to a low scan rate worsen the Kell factor.

Kelvin Scale: Unit of measurement used in TV lighting which indicates the color temperature of a light source.

Kelvin: This is the color temperature scale that takes its name from the scientist Lord Kelvin.

KEM: A brand name for a common "flatbed" film editing machine.

Key Card: A camera card with white lettering on a black background commonly used to add (key in) titles and other written material over a over a background scene.

Key Clip: A control on a video effects unit used to set the threshold level for color or luminance keying.

Key Grip: The chief of grips, often doubling for a construction coordinator and a backup for the camera crew. Key grips work closely with gaffer.

Key Light: The primary illumination for a scene, generally giving the impression of a natural light source.

Key Numbers: Numbers placed on the edge of the film stock by the manufacturer as a unique identification point (35mm - every 16 frames: 16mm - every 20 frames).

Key Picture: A video picture containing the entire content of the image (intraframe encoding), rather than the difference between it and another image (interframe encoding). MPEG I pictures are key pictures. Contrast with delta picture. Also called Key Frame.

Key Source: A video source keyed into background video.

Key: A video special effect in which the level of a video signal is used as a switch which allows selective substitution of picture information from one source with picture information from a different source. Luminance keyers use the amplitude of the monochrome portion of the signal, while chroma keyers use the amplitude of a specific color or hue. Most printed material is inserted into video using luminance keys, while picture information is inserted using chroma keys.

Keyed-in Time Code: As opposed to burned-in time code, SMPTE time code numbers electronically and temporarily keyed over background video. Also called KeyKode.

Keyframe: A set of parameters defining a point in the transition of an effect in a DVE system. For example, a keyframe may define a picture size, position and rotation. Any digital effect must have a minimum of two key frames, start and finish, although more complex moves will use more: maybe as many as 100.

Keying: An special effects process, specific to video, in which an image or text is inserted into another image. See also Chroma Key.

Keystone: The effect of projecting an image onto a surface that is not perpendicular to the axis of the projecting lens. Parallel lines tend to converge in the direction where the surface is closer to the lens.

KHz (Kilohertz): A unit of frequency measurement. One thousand cycles (repetitions) per second or 1000 hertz.

Kick: An object with a shine or reflection on it from another object.

Kicker Light: A light typically placed between the backlight and the fill lights. Sometimes used to simulate the light from a window behind a subject.

Kickoff: The start of production or principal photography.

Kilohertz: See kHz.

Kinescope: A film copy of a television program; made by aiming the film camera at a television screen. Used during the early years of television (before videotape) to record programs that were broadcast live.

Kinetic: Involving motion and movement

Kiss: A light that gently brushes a subject.

KU-band: Signal frequency range between 11 and 14 GHz. that is often used with communications satellites.

L

Lab Roll: A large roll—usually up to 1,000 feet—made up of camera rolls joined together by the lab for printing.

Lag: The tendency in some camera pickup tubes to retain an image after it is no longer presented to the tube. This effect is most evident when a relatively bright image is replaced by a darker field of view and is aggravated when a bright image is television stationary in the field of view for an extended period of time before it is replaced.

Lamp: A reference to the bulb inside a lighting unit, but may sometimes be used to refer to the entire lighting unit.

LANC: See Control-L.

Land: The raised area of an optical disc.

Lap Dissolve: Fading one video or audio source out while simultaneously fading up on (going to) another source. Midway through a lap dissolve both signals will be present in equal proportions.

Lapel Microphone: A personal microphone attached to a lapel.

Laserdisc: A 12-inch (or 8-inch) optical disc that holds analog video (using an FM signal) and both analog and digital (PCM) audio. A precursor to DVD.

Last Frame of Action (LFOA): The last frame of image and sound on a reel. It is important to the people who mix the film (it should be written on the cue sheet), especially if you need to do Pull Ups.

Late Fringe: The time period following prime time (usually after 11PM except in the Central and Mountain Time zones where it starts at 10PM).

Late Order: An order that has been turned in by a salesperson after the log has been processed and accepted.

When the order starts the day of the accepted log, the spots are called "write-ons."

Latency: The time it takes a signal or bit stream to get through a given process such as a compression/ decompression cycle or the delay in data access due to disk rotation.

Latent Image: The image retained by a pickup tube after the source of that image is removed. If a latent image is present on a pickup tube when a camera is turned off, the image may be permanently fixed on the tube.

Latitude: The degree to which a certain film stock can tolerate under- or overexposure. Reversal film, for all practical purposes, has a very little latitude. Color negative has a higher latitude, and particular of its latitude it is tolerant of much more overexposure than underexposure.

Laugh Track: A soundtrack of pre-recorded laughter, usually added in the post-production process to a comedy program with no studio audience.

Lavaliere Microphone: A small microphone often clipped to a performer's tie or shirt.

Layback: Transfer of the finished audio mix back onto the video edit master.

Layer 0: In a dual-layer disc, this is the layer closest to the optical pickup beam and surface of the disc, and the first to be read when scanning from the beginning of the disc's data. Dual-layer discs are 10% less dense than single layer discs due to crosstalk between the layers.

Layer 1: In a dual-layer disc, this is the deeper of the two layers, and the second one to be read when scanning from the beginning of the disc's data.

Layer(ing): A collection, or "pack," of clip layers can be assembled to form a composite layered clip. Layers may be background video or foreground video with its associated matte run. Quantel editing systems allow many layers to be composed simultaneously, so the result can be seen as it is composed.

Layer: The plane of a DVD disc on which information is recorded in a pattern of microscopic pits. Each substrate of a disc can contain one or two layers.

Layoff: Transfer of audio and time code from the video edit master to an audio tape.

Layouts: A person responsible for working out the action before filming begins, including where the characters should be, and the camera angles. Also called Blocking.

Layover: Transfer of audio onto multitrack tape or hard disk. Also referred to as Layup.

LCD (Liquid Crystal Display): A technology used in flat panel displays to show alphanumeric data.

L-C-R-S (Left, Center, Right, Surround): The four playback channels used in 35mm motion pictures, now available on home hi-fi systems. L, C, and R speakers are located behind the screen. The S channel surrounds the audience and may be mono or encoded stereo.

L-cut: An intentional videotape edit where and audio comes before the associated video or the video comes before its corresponding audio. Also called split edit.

Lead (Sound or Video): The sound (or video) of the next footage is introduced before the video (or sound) creating a split edit.

Lead In: The physical area 1.2 mm or wider preceding the data area on a disc. The lead in contains sync sectors and control data including disc keys and other information.

Lead Out: On a single-layer disc or PTP dual-layer disc, the physical area 1.0 mm or wider toward the outside of the disc following the data area. On an OTP dual-layer disc, the physical area 1.2 mm or wider at the inside of the disc following the recorded data area (which is read from the outside toward the inside on the second layer).

Lead Role: The most important character in a movie, often distinguished by gender, e.g. female lead, male lead.

Lead: In news stories, the reporter's opening comments: designed to capture viewer attention.

Leader Numbers: Visible numbers which appear on countdown leader used to cue up a videotape. Normally, numbers from ten to three flashes at one-second intervals. The numbers are generally accompanied by short audio beeps.

Leader: Film footage with visual calibrations, usually in one-second intervals, used to lead into the film proper. Also called Film Leader.

Lead-in: The preceding program on the same network or television station.

Leading Blacks: A term used to describe a technical picture aberration in which the edge preceding a black object appears white.

Leading Lines: In composition the use of either visible or implied lines within a scene to direct the eye to a specific location, generally to the scene's center of interest.

Leadman: Member of the art department who is in charge of swing gangs and/or set dressers and reports to the set decorator.

Lead-out: The following program on the same network or television station.

Lecture Format: An approach to doing a production or program segment which features an individual communicating information directly to a camera—in somewhat the same manner as a teacher addresses a class.

LED (Light Emitting Diode): A light-emitting pixel point used in instrument read-outs and some flat television screens.

Left in Progress (LIP): A program not aired in its entirety due to preemption at the end of the program.

Legacy: A term used to describe a hybrid disc that can be played in both a DVD player and a CD player.

Legal Effects: The lengths for fades and dissolves which can be executed by most printers (16, 24, 32, 48, 64, and 96 frames).

Legs: Of a movie, the continuing to return large box-office figures.

Leko: An ellipsoidal reflector spot light. Usually used for theatrical purposes.

Lens Aperture: See Aperture.

Lens aperture: See F-stop.

Lens Coating: See Coated Lens.

Lens Extender: An optical lens unit to increase the effective focal length of a zoom or fixed focal length lens. Some lens extenders are electrically flipped into position within the lens housing others are manually placed over the front of or behind the lens.

Lens Flare: A bright spot in an image caused by a bright light hitting the elements of a lens. It is caused when light strikes the lens and either causes the entire image to be fogged in appearance, or for a little row of polygons (the silhouette of the iris) to appear from the light hitting the surfaces of the many elements in the lens. It is solved by flagging the lens.

Lens Format: In describing the lens necessary to image on any sensing device, or a solid-state aperture, use common film formats. This will include the following: (a) Hasselblad (2 1/4 by 2 1/4 inch square), (b) 35mm Still (Nikon style), (c) 35mm Motion Picture (Such as PL and/or Panavision mount lenses)

Lens Hood: Also lens shade. A cone or square fastened to the front end of a lens to keep incident light from striking the lens directly and causing lens flare. Letterbox Placing a widescreen image on a conventional TV by placing black bands at the top and bottom of the screen.

Lens Shade: A round or rectangular black hood that fits over the end of a camera lens to shield the lens from strong sidelight or inclement weather. Also called Lens hood.

Lens Speed: The F-stop which transmits the maximum amount of light for a specific lens. The smallest F-stop number.

Lens: An optical device used by a camera to focus an image onto film stock.

Letterbox Filter: Circuitry in a DVD player that reduces the vertical size of anamorphic widescreen video (combining every 4 lines into 3) and adds black mattes at the top and bottom. Also see Filter.

Letterbox: A process by which a widescreen film is presented on video. The top and bottom of the video frame is blackened, and the widescreen film frame is reduced to fit into this frame-within-the-video-frame. Also used to present high definition video on conventional TV sets.

Level Framing: The horizontal edges of the frame are parallel to the horizon.

Level: The strength or amplitude of an audio or video signal.

Levels of Details (LOD): A polygon count saving feature. Several versions of the model are created with several Detail Levels. As the model gets farther away from the viewpoint, the engine switches to a less detailed model.

Lexan: A plastic sheeting material, available in varying widths, that is optically clear and used to protect camera personnel from explosions or the results of other action.

LFAO: See Last Frame of Action.

Libel: A legal term relating to a published (or sometimes a broadcast) statement that tends to subject someone to public ridicule or contempt or to injure the person's reputation.

Library Shot: A shot used in a film, but not originally taken for that film.

Light Level: The intensity of light measured in foot-candles or lux.

Light Meter: A device that measures the amount of light which is either reflected from or is falling on a subject. See Reflected and Incident Light Meters.

Light Pen: A handheld device allowing the user to write or draw on the screen of a cathode ray tube. Highly sensitive light pens are used to read the surface of the screen and to input information.

Lighting Board: A centralized control system for studio lights. Most lighting boards allow for the dimming of lights and include provisions for controlling each AC outlet in a studio lighting grid.

Lighting Cameraman: See Cinematographer.

Lighting Color: The color of a light source, which may be manipulated with gels.

Lighting Combinations Though there are many combination of light direction and source, three of the most common are**:** Three point lighting, High-key lighting, Low-key lighting.

Lighting Crew: A group of technicians who install, operate, and maintain lighting.

Lighting Department: The section of a production's crew responsible for lighting and other electrical matters during filming. Individual positions within this department include: electrician, lighting crew, and genny operator.

Lighting Diffusion: The hardness or softness of a light source. Hard light casts a sharp, definite shadow.

Lighting Direction: The positioning of lights relative to the object being shot. The norm for lighting direction is three-point lighting.

Lighting Director: An individual responsible for the planning, design and set-up of lights for a production.

Lighting Grid: A criss-cross arrangement of pipes suspended below the studio ceiling used to hold lights.

Lighting Intensity: The power of a light source. Regarding the relative intensity of lighting sources, see three-point lighting.

Lighting Plot: A detailed drawing, generally to scale, showing the placement of each light in relation to talent positions and scenic elements.

Lighting Ratio: The relationship between the key and fill lights. Typically 2:1 for color and 3:1 for black and white television.

Lighting Source: See Key Light and Fill Light.

Lighting Tent: A white fabric enclosure that totally covers a subject. Lighting tents are illuminated by two or more soft light sources to produce an extremely flat lighting effect which can be useful in photographing bright, shiny objects.

Lighting: Most productions use artificial lighting when filming for various technical and artistic reasons, both on location or on a set. Lighting is designed by the director of photography in consultation with the director, and is the responsibility of the electrical department.

Lightleak: Stray light that penetrates into a camera giving the film little patches of fog. Also the term for the access point itself. Typically light leaks occur around the camera door or where the magazine is joined to the camera body. Often they can be easily prevented with camera tape around the door.

Lightmap: Pre-rendered "map" of where all the shadows in the scene fall. The lightmap is projected on the polygons to show visual effects of light sources. Because they are pre-rendered, only static objects are included in lightmaps.

Light-Struck Leader: Film which was fully exposed to light which is then used as leader.

Limbo Background: A background of any color or brightness that has no discernible detail.

Limbo Lighting: A style of film lighting which eliminates background light and isolates the subject against a completely dark (or neutral) field.

Limited Effects Theory: A type of communication theory (e.g., social learning theory, vicarious catharsis theory) that regards media as having conditional influences on the viewer; due to intervening variables, the effects of media on the viewer are limited.

Limiter: An amplifier designed to limit or compress signals over a desired level, thus reducing the chances of distortion and keeping the range of signal levels within the range that can be recorded. Unlike an automatic gain control, a limiter does not augment or boost low levels.

Line Compensation: Use of a video line amplifier to pre-compensate for high frequency video signal transmission losses resulting from long distance cable runs (several hundred meters) by boosting those signal frequencies most affected. Without such compensation, deterioration is manifested as loss of fine details and color distortion.

Line Doubler: A set of circuits whose purpose is to paint each scan line on the picture tube twice (three times, four times) and thus fill in the gaps between scan lines. A device (with circuits inside) which converts interlaced video to progressive scan video is also referred to as a line doubler. It too delivers output that has twice as many scan lines per second as its input and it may also use the technique equivalent to painting each scan line twice.

Line Microphone: A highly directional, hand-held microphone, generally mounted within a reinforced foam rubber enclosure and normally used in video field production just out of camera range.

Line Out Monitor: A video monitor displaying the final video output intended for broadcast, recording or distribution.

Line Pair: On a test pattern consisting of closely spaced black parallel lines on a white background, a line pair is one black line together with the white space on one side of it. When photographers say lines of resolution, they mean line pairs.

Line Producer: A producer who is responsible for managing every person and issue during the making of a film. Line producers only work on one film at a time.

Line Store: A line store is a memory used to hold one scan line of video. If the horizontal resolution of the active display is 640 samples and RGB is used as the color space, the line store would have to be 640 locations long by 3 bytes wide. This amounts to one location for each sample and each color. Line stores are typically used in filtering algorithms. For example, a comb filter is made up of one or more line stores.

Line Transformer: An electrical device generally used to change or match the impedance of audio sources. Line transformers are also used to electrically isolate sources of audio.

Linear Audio Track: Audio recording made on a track along the edge of a videotape by television stationary record/play heads.

Linear Editing: The process of editing footage that can only be accessed or played in the sequence recorded. Tape is linear in that it has to be spooled to access any required material and can only play pictures in the order they are recorded. Editing with tape is linear but, although it has been widely used since 1956, the medium was only labeled as linear in the late 1980s to differentiate it from nonlinear systems.

Linear Fader: As opposed to a rotary fader, a volume control device which is moved along a straight line.

Linear PCM: A coded representation of digital data that is not compressed. Linear PCM spreads values evenly across the range from highest to lowest, as opposed to nonlinear PCM which allocates more values to more important frequency ranges.

Linear Perspective: A method of drawing or painting that converts the three dimensions of reality into two dimensions. Originally developed during the European Renaissance, it formed the foundation for how lenses represent a visual field.

Linearity: The ability of a video camera or TV set to accurately reproduce a test pattern with geometric accuracy.

Lined Script: A copy of the shooting script which is prepared by the script supervisor during production to indicate, via notations and vertical lines drawn directly onto the script pages, exactly what coverage has been shot. A given vertical line indicates, via the line's start and end.

Line-level: In contrast to a microphone-level input, a high-level audio input associated with pre-amplified sources of audio.

Line-Locked Clock: A design that ensures that there is always a constant number of samples per scan line, even if the timing of the line changes.

Lineup: A listing of the basic elements in a production in the order in which they will appear.

Lip Sync Error: Situation in motion pictures where picture and sound are not synchronized in time. So named because the error is particularly noticeable as subject lip movement that does not match the words spoken. For both film and videotape, the sound head is several inches away from the projector gate / video heads, and careful adjustment is needed. Video processing units, such as line doublers that accumulate several frames of video in order to do their processing, can introduce lip sync error.

Lip Sync: Having an on-camera performer mouth words to a prerecorded soundtrack to make it appear as if they are actually singing during a production.

LIP: See Left in Progress.

Liquid Gate: A printing system in which the original is immersed in a liquid at the moment of exposure. This is done in order to reduce the adverse effects of surface scratches and abrasions.

Live Action: Film action with living people and real things, rather than action created by animation.

Live Area: A camera's viewfinder actually shows (and records on film stock) a greater area of the scene than will appear in the final product. Markings are etched in the viewfinder to indicate to the camera operator the extents of the "viewable" film (called the live area). An area beyond that (called the safe area) is also marked; it is in this area that the camera operator might direct the boom operator to place the boom microphone.

Live Copy: Copy read by announcer at the time it is aired.

Live Feed: The use of only one transponder to transmit the programming of an ad supported cable network. Hence programming that airs in prime time on the East Coast at 8-11PM will air on the West Coast from 5-8PM. Some of the ad supported cable networks that use a live feed are CNN, ESPN, and The Weather Channel.

Live on Tape: Unedited. Recorded exactly as occurring with no compression of time or alteration of any sequence of events.

Live: Not prerecorded. Occurring at the same instant as it is seen on the television.

Live-Streaming: Feeding (streaming) video/audio or other data to end-users at a specific time (that is, live). Today the term is most often used for on-line services.

LNB: See Low Noise Blockconverter.

Lo/Ro: Left only/right only. Stereo signal (no matrixed surround information). Optional downmixing output in Dolby Digital decoders. Does not change phase, simply folds surround channels forward into Lf and Rf.

Load Resistance: The impedance or resistance (load) that a cable places on a signal being transmitted through it. In the case of a high frequency signal, signal-to-cable matching is essential to prevent signal deterioration. The cable should be terminated by a specific load resistance, usually 50 or 75 ohms. Improper cable loading results in signal distortion, ghost images, color loss and other adverse phenomena. Most video inputs have the proper termination built in.

Loading Booth: A small darkroom sometimes found on a sound stage for loading film into magazines as a roomier alternative to a changing bag.

Local Origination Channel: A channel programmed by a cable system, whether a public access channel, an advertiser-supported entertainment channel or a combination of both.

Locale: See Regional Code.

Location Filming: Filming which occurs at a place not constructed specifically for the production. Typically this is either outdoors, a well-known location, or a real place which suffices. Also called On Location.

Location Manager: A person who manages various aspects of filming on location, such as arranging with authorities for permission to shoot in specific locations.

Location Mixer: A sound mixer responsible for mixing sounds recorded on location.

Location Scout: A person who looks for suitable locations for filming.

Location Settings: Pre-existing settings that are chosen as backgrounds for television programs.

Location Sketch: A rough drawing of a setting for a remote telecast.

Location Sound: This is the sync sound, or any other sort of wild track or room tone that was recorded at the shoot. Same as Production Sound.

Location Survey: Written appraisal of production requirements required for a field production.

Lock it Down: A direction given by the assistant director for everyone on the set to be quiet. It is called just prior to speed.

Locked Cut: The so-called final cut of a film when there are to be no more changes to picture.

Locked Down Shot: A shot taken with the pan and tilt releases on the tripod tightened so that the camera will not move. Often done for certain effects where camera movement would ruin the illusion, such as a cut that causes a character to magically disappear from a scene.

Locked: When a PLL is accurately producing timing that is precisely lined up with the timing of the incoming video source, the PLL is said to be "locked." When a PLL is locked, the PLL is stable and there is minimum jitter in the generated sample clock.

Locks & Limits (Parental Controls): Allows you to restrict viewing of rated movies (based on the motion picture rating system) or to lock out entire channels.

Lock-up Time: The time required by a tape machine to stabilize after it has been rolled.

LOD: See Levels of Detail.

Log: An operational document for broadcast operations, generally issued daily, which includes information such as program source or origin, scheduled program time, program title and other program-related information.

Logging a Tape: The listing of time-code numbers of video and audio segments prior to editing.

Logging: Making a list of shots. This usually applies to recorded tape or film and can be vital for speeding the selection of material for editing. Especially true when going to a disk-based editing system where digitizing time can be cut to a minimum. The list, or log, may be on paper or on computer disk. With the latter it can be used directly to control the loading of footage to disk.

Logical sequence: An editing approach in which segments are assembled in a natural, time-based progression.

Logical unit: A physical or virtual peripheral device, such as a DVD-ROM drive.

Logical: An artificial structure or organization of information created for convenience of access or reference, usually different from the physical structure or organization. For example, the application specifications of DVD (the way information is organized and stored) are logical formats.

Logo: Symbol or trademark that represents a company, or organization.

Long Lens: A lens with a focal length greater than 25mm in 16mm, or 50mm in 35mm, which, like binoculars, will provide a view that magnifies a small area.

Long shot (LS): A shot taken from a considerable distance. Often the LS serves as an establishing shot. (i.e., a human figure taken so it is shorter than the height of the screen).

Long-Form TV: Also known as "MOWs," these are movies and miniseries that are aired on free or pay television.

Longform: Longer-length programs, not shortform. Editing longform programs involves handling large amounts of material and so requires a large editing store and the management to handle the quantity of material efficiently. Much of the material will be edited by cuts only and much of the basic editing detail may be defined by an EDL from off-line. The first process in on-line nonlinear would then be to conform the edit. There may also be some complex work involved. See also Shortform.

Longitudinal Recording: To record audio signals on tape along its longitudinal tracks. Sometimes called LNG Recording.

Longitudinal Time Code: In contrast to VITC, a method of recording digital SMPTE/EBU time-code information along the edge of a videotape, generally on an audio track.

Looking Space: In composition the blank space left in front of a subject looking off screen. Also referred to as leading the subject.

Loop: Slack film above and below the gate to allow a transition from the constant motion of the supply and take up rollers to the intermittent motion that takes place at the gate. A Loop is also a small magnifier useful in the editing room.

Looping: A term used to describe the chaining of a video signal through several video devices (distribution amplifiers, VCRs, monitors, etc.). A VCR may be hooked up to a distribution amplifier which is supplied with a video input connector and a loop output connector. When a signal is fed to the distribution amplifier, it is also fed unprocessed to the loop output connector (parallel connection) on the distribution amplifier. In turn, the same signal is fed to another device which is attached to the first one and so on. Thus a very large number of VCRs or other video devices can be looped together for multiple processing. Looping is also the process of re-recording dialogue in postproduction to correspond with dialogue that was originally part of a scene. Often sound is looped to improve quality and to eliminate distracting sounds in the original recording.

Lossless Compression: Compression techniques that allow the original data to be recreated without loss.

Lossy Compression: The data compression technique of shrinking a media file by eliminating perceptually insignificant information. Since lossy compression introduces inaccuracies, it should only be used with graphics, audio, and video. Data files and executable programs can only be compressed with a lossless algorithm (i.e., a zip file). Lossy image compression is the exact opposite of lossless. The regenerated image is different from the original image. The differences may or may not be noticeable, but if the two images are not identical, the compression was lossy.

Loudness Meter: A volume units meter which responds to the loudness of sounds as humanly perceived, as opposed to only the milliwatt DB reference used by most VU meters.

Loudness: The perceived strength or intensity of a sound.

Low Angle: A shot in which the camera is lower than the filmed object; thus the camera looks up at the actor/object.

Low Band: Video recording system that uses relatively low FM carrier frequencies. Often used in non-professional videocassette systems.

Low Con Print: A low contrast print specifically for transfer to video, which favors less contrast in the transfer process.

Low Contrast Filter: An optical camera filter that lowers contrast of subject matter and mutes colors. Often used to bring the brightness ratio of a scene within video range and allow more detail in dense shadow areas.

Low Contrast Original: An original reversal film which is designed to yield prints having good projection contrast.

Low Noise Blockconverter (LNB): A microwave receiver circuit that lowers the microwave frequencies, allowing them pass easily through a coax wire to a satellite receiver.

Low Pass Filter: An electrical filter that passes frequencies below a specified frequency and attenuates those above. Required in digital audio to prevent sampling of signals above one-half the sampling frequency.

Lowboy: A heavy duty rolling stand, usually with a combo head, but without the height of a "highboy."

Lower-third: the bottom third of the TV screen, where titles of interview subjects and other text are most often super-imposed over video.

Lowest Effective Rate: A complex set of rules by the FCC that require television stations to provide commercial air time to political candidates at rates no higher than the television station's "most favored" advertiser. The effect of bonus spots, SPINS, ROS, "deal-makers," and other cost-reducing tricks must be factored into the calculation for each "class" of airtime. Also, if a lower effective rate is offered to an advertiser during the period leading up to the election but is not actually booked, the offered rate is still considered available to a candidate.

Low-frequency Roll-off: An audio circuit associated with microphones and amplifiers which attenuates undesirable low frequencies.

Low-key Lighting: Lighting characterized by a high key-to-fill ratio which results in predominant shadow areas. Typically used for night scenes in dramatic productions.

Low-pass Filter: A filter that attenuates frequencies above a specified frequency and allows those below that point to pass.

Low-power TV (LPTV): A classification of TV television station designed to serve a small community through the use of a transmitter with very limited power.

LP: Long-playing record. An audio recording on a plastic platter turning at 331/3 rpm and read by a stylus.

LPCM: See Linear PCM.

LPTV: Low-power TV.

LS: See Long Shot.

L_t/R_t: Left total/right total. Four surround channels matrixed into two channels. Mandatory downmixing output in Dolby Digital decoders.

LTC (Linear Time Code): Time code recorded on a linear analog track on a videotape.

LTC (Longitudinal Time Code): SMPTE (Society of Motion Picture and Television Engineers) time code standard usually recorded onto the linear It is recorded along the tape in the forward direction of tape run. A VTR cannot reproduce LTC when tape run stops to output a still picture. The output level of LTC is very low when tape runs slowly. So an LTC read error is likely to occur while the VTR is playing back in slow motion. See also Time code and VITC.

Luma: Video systems use a signal that has two pieces: the black and white part, and the color part. The black and white part is the luma. It was the luma component that allowed color TV broadcasts to be received by black and white TVs and still remain viewable.

Lumen: Measurement of light quantity. Lumens per square foot equals foot candles.

Luminance Channel: The black and white circuitry associated with a color video signal. See Luminance.

Luminance Key: An electronic key effect controlled by the brightness of one video source.

Luminance Noise: Noise which manifests itself in a video picture as white snow, typically caused by one of the following situations: 1. Low signal level due to poor lighting conditions 2. Poor video signal processing 3. Low quality videotapes 4. Excessively long video cables used without pre-compensation 5. Dirt on the video recorder heads which interferes with reading and writing 6. Over-enhancement of the video signal.

Luminance Signal: A signal that determines the brightness of the picture. Also called Y Signal.

Luminance: The degree of brightness (black and white portion of the video signal) at any given point in the video image. A video signal is comprised of luminance, chrominance (color information) and sync. If luminance is high, the picture is bright and if low the picture is dark. Changing the chrominance does not affect the brightness of the picture.

Lux: Unit of light intensity used in metric countries. One foot-candle is equal to about 10.7 lux.

Lyrical: A stylistic approach which emphasizes the romantic, thereby producing in the viewer intense emotional feelings.

Lyricist: A writer of song lyrics.

Lyrics: The words sung in a song; also refers to their writer.

M

M Byte: One million (10^6) bytes. Not to be confused with MB or megabyte (2^{20} bytes).

M&E (Music and Effects): After a mix a big production will have an M&E track made, which is used when the film is dubbed into other languages so that all the Music and Effects do not also have to be redone. An M&E track is only essential if you plan on dubbing the film into a different language.

Macguffin: Term used by Alfred Hitchcock to refer to an item, event, or piece of knowledge that the characters in a film consider extremely important, but which the audience either doesn't know of or doesn't care about.

Machine Leader: Strong leader threaded through a film processing machine which is used to pull film through the machine during its operation.

Macro Lens: A lens that can be used for extremely close to the subject. The focusing ring will keep going past the lowest setting (on the Switar lens a red ring will appear to let you know) all the way around again. When in macro the distances on the focusing ring no longer apply.

Macro Setting: A zoom lens setting that allows for extreme close-ups of subjects.

Macroblock: In MPEG MP@ML, the four 8x8 blocks of luma information and two 8x8 blocks of chroma information form a 16x16 area of a video frame.

Macrovision: An anti-taping process that modifies a signal so that it appears unchanged on most televisions but is distorted and unwatchable when played back from a videotape recording. Macrovision takes advantage of characteristics of AGC circuits and burst decoder circuits in VCRs to interfere with the recording process.

Mag Stock: Magnetic sound recording stock which has

edge perforations that match those perfs. on the picture stock, thereby allowing it to be pulled along with the picture at the same speed and relative position. Also called Mag Track or Magnetic Film.

Mag Track: A piece of film that is coated with an emulsion of magnetic oxide instead of silver halides. Basically, it is sound recording tape that is the same size as film, complete with perforations. For editing, all the sound, location sound and additional sound, is transferred to mag stock, where it is run on an editing machine in tandem with picture, one frame of picture equaling one frame of sound.

Magazine: An attachment to a camera with one or two light-proof chambers that hold 400 or 1,000 feet of film. One camera will typically have two or three magazines which can be loaded ahead of time.

Magic Hour: Magic Hour is the time just before sunrise or the time just after sunset. This is when there is enough light to get an exposure on buildings and on various landscapes but car headlights and building lights can still be seen as on.

Magnetic Soundtrack: A composite print in which the soundtrack is recorded on the attached strip of magnetic tape. Largely obsolete due to high cost and maintenance difficulties. Also called Magnetic Print.

Magnetic Tape: A ribbon of plastic with a coating on it that is sensitive to magnetic impulses created by electricity. In analog technology, these magnetic impulses are modulated on the tape in a fashion parallel to the sound wave's modulation. In digital technology, magnetic tape is used to record sounds encoded as a string of numbers that will later be converted into sound.

Magneto-optical: Recordable disc technology using a laser to heat spots that are altered by a magnetic field. Other formats include dye-sublimation and phase-change.

Main Data: User data portion of each sector. 2048 bytes.

Main Level (ML): A range of proscribed picture parameters defined by the MPEG-2 video standard, with maximum resolution equivalent to ITU-R BT.601 (720x576 x 30). See also Level.

Main Profile (MP): A subset of the syntax of the MPEG-2 video standard designed to be supported over a large range of mainstream applications such as digital cable TV, DVD, and digital satellite transmission. See also Profile.

Maintenance Engineer: A person responsible for general maintenance and repair.

Major Market: One of the 50 largest metropolitan areas in numbers of households.

Majors: The major Hollywood movie producer/distributor studios (MGM/UA, 20th Century Fox, Sony Pictures, Warner Bros, Paramount Pictures, Universal, and Disney).

Make Good: An announcement offered to or accepted by an advertiser to replace a commercial that did not run as scheduled or was aired improperly.

Malice: Untrue statements, often motivated by ill-will or an attempt to harm another. When used in the legal sense, malice can refer to a published or broadcast statement that results from a careless disregard for the truth.

Management: The process of organizing digital media such as indexing, data entry, watermarking and rights management for effective distribution of that content.

Manifest Content: In a content analysis of a television text, the characters and their actions.

Mark it!: What to say to the person with the slate to get them to clap the sticks together.

Mark: The clapping of the clapstick to create a Sync Mark for the shot. A "Mark" might also be a piece of tape on the floor that indicates where an actor should stand.

Markers: Points inserted to identify areas of interest on recorded footage for future quick access and reference. These may be inserted either while replaying or "live" as the material is recorded. See also Scene Select.

Marquee Value: The projected value of a film before its released. Gauged according to the reputations of the major stars and creative personnel. Merchandising Non-filmic goods which are released under the film's theme, such as toys, T-shirts and posters.

Married Print: A positive print which carries both picture and sound on it. Sometimes called a composite print.

Mask: A covering of some type placed before the camera lens to block off part of the photographed image. A mask can also refer to an aperture plate inserted behind a projector lens in order to obtain a desired aspect ratio. In video, letterboxing is an electronic form of a mask..

Master Black Control: An adjustment on camera CCUs which controls the blanking, pedestal or black level in the video.

Master Control Switcher: Video switcher with limited video effect capabilities designed to serve the specific needs of a master control area. Most such switchers control video and audio simultaneously.

Master Control: An audio and video control center which is the final switching point before signals are sent to the television transmitter.

Master Scene Script Style: An approach to script writing where only the basic scenes are described. Decisions on the various shots within the scenes are not outlined in the script and are left to the discretion of the director.

Master Shot: A single shot, usually a wide shot, that incorporates the whole scene from beginning to end. Typically a master shot will be filmed first, and then all the close-ups and other shots afterwards.

Master: The final result of the edit. Until now this has been in the form of a tape but, with disk-based broadcasting, it could exist as original footage and edit decisions held on a server. The first-generation tape edited from raw footage.

Mastering: The process of replicating optical discs by injecting liquid plastic into a mold containing a master. Often used inaccurately to referred to as premastering.

Match Cut: A cut made on action or movement between two shots in which the action has been overlapped either by repetition of the action or by the use of more than one camera.

Match Dissolve: A dissolve linking images which have similar content.

Match Frame (edit): An edit in which the last frame of the outgoing shot is in sync with the first frame of the incoming shot, such that the incoming shot is an extension of the outgoing.

Match on Action: A juxtaposition of shots that carries a movement across the break between two shots.

Matching Transformer: A impedance-changing device used to make different pieces of equipment electronically compatible. Often used to alter the output of high-impedance microphones so they can be used with low-impedance equipment inputs.

Matching: Arranging for the impedances presented by a load to be equal to the internal impedance of the generator. This is essential to avoid loss of power. In microphones, the loss results in poorer signal-to-noise ratio. Matching is done by means of a transformer.

Mater: A small adjustable clamp with a baby stud which can be interchanged with a variety of accessories.

Materials Contract: Contract for representation by an agency with regard to the sale of a work that the writer has created on his own, in a situation where the writer was not hired to create the work.

Matrix Encoding: The technique of combining additional surround-sound channels into a conventional stereo signal. Also see Dolby Surround.

Matrix Switcher: A device which uses an array of

electronic switches to route a number of audio/video signals to one or more outputs in almost any combination. Production quality matrix switchers perform vertical interval switching for interference free switching. Matrix switchers may be operated with RS-232 or RS-422 controls, enhancing flexibility.

Matte Artist: A person who creates artwork (usually for the background of a shot) which is included in the movie either via a matte shot or optical printing.

Matte Box: A square shade that goes in front of the lens, usually supported by a pair of rods that attach to the camera. A matte box often has filter holders for square glass filters.

Matte Key: Commonly refers to keying letters or symbols from one video source into background video. In the process the letters or symbols can be electronically filled in with any desired color.

Matte Shot: A process of combining two or more separate shots onto a single print so that the composite image looks as though it had been photographed normally. In a traveling matte shot the contours of the blacked-out areas of each shot are varied from frame to frame to create a composite image that can pan or track.

Matte: A means of covering up part of an image. In photography and video it (as plural) generally refers to the aperture plates or other means of hiding from view the top and bottom edges, leaving what will ultimately be a widescreen movie picture. Whereas without the mattes the displayed image would be taller with extra material at the top and bottom that is really not part of the presentation.

Maxi-Brute: A 9 light unit with (9) 1000 watt PAR 64 lights.

MB (Megabyte): 1 million bytes.

MBPS (Megabits/second): Millions (10^6) of bits per second.

MCU: See Medium Close Up.

ME Track: This refers to the music and effects tracks which are combined into one (or a stereo pair) for use with foreign language re recording of a film or video program.

Meat Axe: A grip arm-like accessory which is designed to clamp onto the hand rail of a studio overhead catwalk, or other suitable surface, and has a gobo head at the end of the arm.

Media Instructions: Instructions from an advertising agency as to what tapes or films to run for a particular spot schedule.

Media Inventory: A listing of the tapes and films on hand at a television station, usually with the shelf locations in storage cabinets.

Media Key Block (MKB): Set of keys used in CPPM and CPRM for authenticating players.

Media Text: Any item in the mass media (e.g., a TV commercial or program, film, magazine, interview, public appearance, etc.).

Media: The tapes or films on which spots are supplied to the television station.

Mediamark Research Inc. (MRI): A research company that publishes biannual syndicated reports measuring the cumulative viewing of cable networks, product usage, magazine readership, etc. MRI's software package is called MEMRI.

Medium Close Up (MCU): A shot between a MS and a CU. (i.e., a human figure taken from the chest up).

Medium Long Shot (MLS): A framing in between long shot and medium shot.

Medium Shot (MS): A camera angle often used to describe a shot of character from the waist up.

Megapixel: A term referring to an image or display format with a resolution of approximately 1 million pixels.

Melodrama: A play or film based on a romantic plot and developed sensationally, with little regard for convincing motivation and with strong appeal to the emotions of the audience.

Memory Effect: Power-loss phenomenon alleged of NiCad camcorder batteries, attributed to precisely repetitive partial discharge followed by complete recharge, or long-term overcharge. Considered misnomer for "voltage depression" and "cell imbalance."

Memory: Data storage used by computers or other digital electronics systems. Read-only memory (ROM) permanently stores data or software program instructions. New data cannot be written to ROM. Random-access memory (RAM) temporarily stores data-including digital audio and video-while it is being manipulated, and holds software application programs while they are being executed. Data can be read from and written to RAM. Other long-term memory includes hard disks, floppy disks, digital CD formats (CD-ROM, CD-R, and CD-RW), and DVD formats (DVD-ROM, DVD-R, and DVD-RAM).

Menu: In DVD-Video there are two kinds of menus, System Menus and Interactive Menus. There are six types of System Menus: Title Menu, Root Menu, Audio Menu, Sub-picture Menu, Angle Menu and PTT or Chapter Menu.

Mercer Clip: A trade name for a small plastic clip which is used to hold film ends together during film assembly.

Meridian Lossless Packing (MLP): A lossless compression technique (used by DVD-Audio) that removes redundancy from PCM audio signals to achieve a compression ratio of about 2:1 while allowing the signal to be perfectly recreated by the MLP decoder.

Mesecam: A technique of recording SECAM video. Instead of dividing the FM color subcarrier by four and then multiplying back up on playback, MESECAM uses the same heterodyne conversion as PAL.

Meta Data: Data describing other data or information

about the asset. Metadata can include but is not limited to, information about how the asset was acquired, who holds the rights to it, pointers to the related contracts, its location, format and descriptions of its contents.

Metal Evaporated Tape: A high-quality audio and videotape which can record higher frequencies with a high signal-to-noise ratio.

Metal Oxide Semiconductor (MOIS): A solid-state electronic imaging device used in some video cameras.

Metal Tape: Magnetic tape coated with fine metallic particles, noted for its high recording density.

Metaphor: A comparison between two otherwise unlike entities, usually achieved in films by montage.

Metered Markets: A local market television audience measurement service which, at the present time, tabulates the household delivery daily in 32 major DMA's covering 55% of the country.

Method Acting: A style of acting formalized by Konstantin Stanislavsky which is believed by some to create more realistic performances. Essentially, the theory requires actors to draw experiences from their own personal lives that correlate to the character they are playing: an extremely demanding process emotionally. In some cases, "method" actors take the theory even further by arranging events in their private lives to resemble the lives of their characters. See the trivia entries for Down and Out in Beverly Hills and One Flew Over the Cuckoo's Nest for examples, and the trivia entry for Marathon Man for an amusing anecdote.

Method: Naturalist performance style which encourages the actor to become the character, at which point the gestures/dialects necessary for the performance will emerge organically; approaches used to achieve this union between actor and character are emotional memory, sense memory, and improvisation.

Metteur en Scène: The director who creates the mise en scène.

Mezzanine Level Compression: Compression system that provides contribution quality.

MHz (Megahertz): One million cycles per second.

Mickey Mousing: Creating music that mimics or reproduces a film's visual action, as, for example, in many Walt Disney cartoons.

Mickey: An open faced 1K lighting unit. Also known as a "Redhead."

Microphone Impedance: The nominal load impedance for a microphone indicates the optimum matching load which utilizes the microphone's characteristics to the fullest extent. Impedance is a combination of dc resistance, inductance and capacitance, which act as resistances in ac circuits. An inductive impedance increases with frequency; a capacitative impedance decreases with frequency. Either type introduces change in phase.

Microphone Preamplifier: A microphone is a device which converts sound waves to electrical impulses (transducer). Microphones typically generate very low signal levels requiring low noise, high fidelity, pre-amplification to boost the output signal to a level compatible with audio amplifier circuitry. Good microphone preamplifiers provide precise matching of microphone impedance and low noise electronic components.

Microphone: A device which converts sound into electrical impulses, usually for recording or amplification.

Microprocessor: A single computer chip which is capable of processing information according to programmed instructions. Microprocessors are used in digital cameras, for example, to set up and maintain a camera's optimal performance.

Microwave Relay: A point-to-point, line of sight transmission method which relies on high frequency RF energy as a medium. Often a number of microwave relays are spaced 50 or so miles apart to send signals over great distances.

Microwave: A directional, point-to-point transmission, generally of an audio and video signal using high-frequency RF energy. Typically used to get a TV signal from the field to the studio or from the studio to the transmitter.

Middle Area: Unused physical area that marks the transition from layer 0 to layer 1. Middle Area only exists in dual layer discs where the tracks of each layer are in opposite directions. 918)

MIDI (Musical Instrument Digital Interface): System of communication between digital electronic instruments allowing synchronization and distribution of musical information.

Mid-side Microphone: A stereo microphone technique commonly used in broadcast television. Important centered sounds are sharper and more stable than those produced by the X-Y method. Also called M-S Microphone.

MII: Portable, professional video component camera/ recorder format, utilizing 1/2" metal particle videotape.

Millennium Group: The group of companies proposing the Galaxy watermarking format. (Macrovision, Philips, Digimarc)

Miniatures: Also model or miniature shots models photographed to give the illusion that they are full scale objects; for, example, the space craft in Star Wars were all models on wire guides.

Minicam: A small, lightweight television camera.

MiniDV: This is the most affordable, widely used tape format in the digital field. Each tape, about the size of a stack of 30 business cards, can record 1 hour of high quality video and audio.

Mini-Majors: Studios which are large but not as large as the majors, Embassy, Gramercy, etc.

Minimalism: Auster filmmaking in which the cinematic elements are keep to an absolute minimum.

Minimum Basic Agreement: Fees and basic working conditions for the employment of writers within the entertainment industry as negotiated and set forth by the Writers Guild of America.

Miniseries: A television series with a set number of episodes which tell a complete story, usually filmed at the same time.

Minus Lens: A supplementary lens that increases the effective focal length of a basic lens to produce telephoto effects.

MipMap: Similar to LODs, but for textures. MipMaps are a series generated from a base texture, each with decreased size. As a polygon gets farther away from the viewpoint, the MipMap switches to a lower-sized texture. This can increase drawing speed and avoids aliasing and moire patterns.

Mise en Scene: Literally translated as "what's put into the scene," this is the sum total of all factors affecting the artistic "look" or "feel" of a shot or scene. These can include shot selection, shot composition, production design and set decoration.

Mix Master: This is a copy of the sound mix on mag stock, or on DAT, which you sometimes have to request in addition to the optical track. It is always a good idea to get a copy of the mix on tape, which will be of much better quality than the optical track for transfer to video, or to save some mixing time in the event you have to remix.

Mix: Electrically combining the signals from microphones, tape, and/or reproducers and other sources.

Mix: The combining of different sounds onto the final soundtrack. Movie of the Week A prime-time position for the screening of a movie on television. Net Profit The amount of excess money available after exhibitors, distributors, cast, crew and financiers have taken their share.

Mixdown: Performing the mixes that have been set up, usually applies to audio. Many tracks and associated fader settings may be used to create the desired result but, at some point, this must be committed and reduced to, typically, a single stereo channel. Mixdown is also used where yet more tracks need to be added—beyond the number that can be handled simultaneously.

Mixed Mode: A type of CD containing both Red Book audio and Yellow Book computer data tracks.

Mixed-sources: Two or more light sources with significantly different color temperatures. Commonly, sunlight and incandescent light.

Mix-effects Bank: A double row of buttons on a switcher which can be used in the creation of video effects.

Mix-effects Monitors: Video monitor dedicated to setting up and viewing video effects from one or more mix-effects banks.

Mixer: A device that blends two or more audio or video sources.

Mixing House: A sound studio specifically for mixing sound for film.

M-JPEG: See Motion JPEG.

MKB: See Media Key Block.

MLP: See Meridian Lossless Packing.

MLS: See Medium Long Shot.

MMDS: See Multichannel Multipoint Distribution Service.

MO: Magneto-optical rewritable discs.

Mobile Unit: A mobile television control room used to cover events in the field. Signals are commonly sent back to the studio by microwave or satellite.

Mode of Production: An aesthetic style of shooting that relies upon a particular technology and is governed by a certain economic system. Television's two principal modes of production are single-camera and multiple-camera.

Mode of Representation: Manner in which a non-narrative television program depicts historical reality and addresses itself to the viewer about that version of reality; modes include expository, interactive, observational, and reflexive.

Model Release: Agreement to be signed by anyone appearing in a video work, protecting videomaker from right of privacy lawsuit. Specifies event, date, compensation provisions, and rights being waived.

Modeling Effect: Effectively using light to create depth and dimension in subject matter.

Modification Level: An identification of a software release. Each subsequent modification level contains those enhancements and bug fixes that have been developed most recently by the software supplier.

Modular Camera: A video camera that can be configured in various ways in terms of the attachment of different viewfinders, recorders, etc.

Modulation: The superimposition of audio and video signals on a radio frequency carrier wave so that the signals may be broadcast.

Modulator: A circuit that combines two different signals in such a way that they can be pulled apart later.

Moiré Effect: Color reverberations that occur in video when narrow, contrasting stripes of a design interfere with the scanning lines.

MOIS: See Metal Oxide Semiconductor.

Monaural: Single-channel audio as opposed to stereo audio.

Monitor: A TV set, normally without an RF tuner or audio circuitry, used for the checking of color, composition, etc., during a production.

Monochromatic Stock: The earliest film stock, sensitive only to the blue area of the color spectrum but capable of great depth of field. Monochromatic film was gradually eclipsed by orthochromatic stock starting in 1918.

Monochrome Signal: A monochrome signal is a video source having only one component. Although usually meant to be the luma (or black-and-white) video signal, the red video signal coming into the back of a computer display is monochrome because it only has one component.

Monochrome: A black and white picture, as opposed to color.

Monologue: A character speaking alone on screen or, without appearing to speak, articulating her or his thoughts in voice-over as an interior monologue.

Monopod: One-legged camera support. See Tripod.

Monotonic: This is a term that is used to describe ADCs and DACs. An ADC or DAC is said to be monotonic if for every increase in input signal, the output increases.

Montage Editing: Although the term as used in early film work had a different meaning, today montage editing refers to a rapid, impressionistic sequence of disconnected scenes linked by a variety of transition devices that are designed to communicate feelings or experiences. A montage does not tell a story by developing an idea in a logical sequence.

Montage: As used in film aesthetics, montage refers to a rhetorical juxtaposition of shots arranged so that the clash between two adjoining images creates a third, independent entity and a whole new meaning. This idea is associated with the work and theory of Sergei Eisenstein, who identified five levels, or types of montage: metric, rhythmic, tonal, overtonal, and intellectual.

Mood Music: Background music intended to evoke a particular feeling.

Morphing: A visual—generally computer-generated—effect in which a person or object is progressively transformed into another person or object.

MOS: A shot, a sequence, or a film that is shot without sound, which is added later. MOS stands for "Mit Out Sound," and derives from an old Hollywood story about a German director asking for a shot to be filmed "mit out sound," and the camera assistant complying with this request by writing "M.O.S." on the slate.

Mosaic Filter: An array of colored filters that cover a CCD chip used to separate a picture into its primary color components.

Mosaic: Visual effect in which an image is broken down into a moving pattern of squares.

Mosquitoes: A term referring to the fuzzy dots that can appear around sharp edges (high spatial frequencies) after video compression. Also known as the Gibbs Effect.

Mother: The metal disc produced from mirror images of the father disc in the replication process. Mothers are used to make stampers, often called sons.

Motif: A recurring subject, theme, or image in a film.

Motion Adaptive: A design that senses motion in order to alter the way it functions, for the purpose of avoiding or reducing some motion related artifacts.

Motion Artifacts: Picture defects that appear only when there is motion in the scene. This often occurs with extreme signal compression or when international broadcast standards are converted.

Motion Blur: Shots of objects that quickly move in the camera's frame, and/or shots with a slow shutter speed are likely to produce a "smearing" effect, since the object is in a range of positions during a single exposure. Motion blur is an usually considered a desirable artifact that is most inherent in motion picture photography where the camera shutter is open for a period of 1/48 of a second and objects in the image continue to move during that period.

Motion Capture: An animation technique in which the actions of an animated object are derived automatically from the motion of a real-world actor or object. See also Rotoscoping.

Motion Compensation: Image processing system that uses motion detection, estimation and adaptation in order to produce smoother temporal image sequences. Typically used in standards conversion and video compression.

Motion Control: A camera setup which records the motion of a camera during a shot so that visual effects can be easily synchronized with the photographed scene.

Motion Detection: Operation of video signal processing aimed to produce an output indicating the pixels or groups of pixels which belong to moving objects as opposed to static portions of the picture.

Motion Estimation: In video encoding, the process of analyzing previous or future frames to identify blocks that have not changed or have only changed location. Motion vectors are then stored in place of the blocks. This is very computation-intensive and can cause visual artifacts when subject to errors.

Motion JPEG: JPEG compression or decompression that is applied real-time to video. Each field or frame of video is individually processed.

Motion Picture Association (MPA): The Motion Picture Association of America and its international counterpart, the Motion Picture Association serve as the voice and advocate of the American motion picture, home video and television industries, domestically through the MPAA and internationally through the MPA. Before 1994, the MPA was known as the Motion Picture Export Association of America.

Motion Picture Editors Guild: A professional union for picture and sound editors, which now also includes re-recording mixers, projectionists, recordists, microphone boom operators, engineers, and story analysts.

Motion Picture: Movie, Film, Flick, Picture.

Motion Vector: A two-component video signal showing the magnitude and direction of moving object displacement over given time interval, e.g. during one TV field. Usually represented in Cartesian coordinates, but, could equally well be represented in polar notation.

Motion-caption Device: A system by which the movement of three-dimensional objects or humans is traced by a computer.

Motivated Lighting: A lighting style in which the light sources imitate existing sources, such as lamps or windows.

Motivation: In narrative structure, a catalyst that starts the story's progression—a reason for the story to begin (usually a character's lack or desire).

Moving Pictures Experts Group (MPEG): The organization which defined the standards for moving pictures, like movies.

Moving-coil Microphone: Microphone which depends upon a diaphragm connected to a moving coil suspended in a magnetic field.

Moviola: Trade name of an American editing machine.

Moviscop: Spelled Moviscop but pronounced "movie-scope." This is a small, 16mm table-top viewer, often used on an editing bench.

MOW (Movie of the Week): Industry shorthand for any film produced specifically for television and not shown initially in theaters.

MP@ML (Main Profile at Main Level): The common MPEG-2 format used by DVD (along with SP@SL).

MP3 (MPEG-1 Layer III Audio): A perceptual audio coding algorithm. Not supported in DVD-Video or DVD-Audio formats.

MPA: See Motion Picture Association.

MPEG 1: MPEG-1 was the first MPEG standard defining the compression format for real-time audio and video. The video resolution is typically 352 x 240 or 352 x 288, although higher resolutions are supported. The maximum bit rate is about 1.5 Mbps. MPEG-1 is used for the Video CD format.

MPEG 2 Video: Video encoded in accordance with the ISO/ IEC 13818 specification.

MPEG 2: Compression standards for moving images and audio are set by the Motion Pictures Expert Group (MPEG), an international committee of industry experts. MPEG-2 is the basis for ATSC digital television transmission in the U.S.

MPEG 3: MPEG-3 was originally targeted for HDTV applications. This was incorporated into MPEG 2, so there is no MPEG 3 standard.

MPEG 4: MPEG-4 uses an object-based approach, where scenes are modeled as compositions of objects, both natural and synthetic, with which the user may interact. Visual objects in a scene are described mathematically and given a position in a two- or three-dimensional space. Similarly, audio objects are placed in a sound space. Thus, the video or audio object need only be defined once; the viewer can change his viewing position, and the calculations to update the audio and video are done locally. Classical "rectangular" video, as from a camera, is one of the visual objects defined in the standard. In addition, there is the ability to map images onto computer-generated shapes, and a text-to-speech interface.

MPEG 7: MPEG-7 standardizes the description of multimedia material (referred to as metadata), such as still pictures, audio, and video, regardless if locally stored, in a remote database, or broadcast. Examples are finding a scene in a movie, finding a song in a database, or selecting a broadcast channel. The searcher for an image can use a sketch or a general description. Music can be found using a "query by humming" format.

MPEG Audio Layer 3 (MP3): CODEC for audio compression capable of 10: 1 compression with no noticeable loss in quality.

MPEG Audio: Audio compressed according to the MPEG perceptual encoding system. MPEG-1 audio provides two channels, which can be in Dolby Surround format. MPEG-2 audio adds data to provide discrete multichannel audio. Stereo MPEG audio is the mandatory audio compression system for 625/50 (PAL/SECAM) DVD-Video.

MPEG I/II Compression: Method to compress digital signals. Thanks to compression it is possible to combine several programs into one satellite transponder.

MPEG Video: Video compressed according to the MPEG encoding system. MPEG-1 is typically used for low data rate video such as on a Video CD. MPEG-2 is used for higher-quality video, especially interlaced video, such as on DVD or HDTV.

MPEG: See Moving Pictures Experts Group.

MS: See Medium Shot.

MSO: See Multiple System Operator.

Mt. Fuji: See SFF 8090.

MTBF (Mean Time Between Failure): A measure of reliability for electronic equipment, usually determined in benchmark testing. The higher the MTBF, the more reliable the hardware.

MTS (Multichannel Television Sound): A generic name for various stereo audio implementations, such as BTSC and Zweiton.

Multi_PGC Title: In DVD-Video, a Title within a Video Title Set (VTS) that contains more than one Program Chain (PGC).

Multicasting: In digital TV, individual broadcast television stations may simultaneously transmit four or more programs. Because digital television allows you to pack much more information into the allotted signal, we can transmit multiple channels on the same bandwidth instead of just one.

Multichannel Multipoint Distribution Service (MMDS): A wireless distribution service for cable-type programming or data.

Multichannel: In film, used to refer to a final mix that includes more than stereo information (i.e., LCRS or six-channel surround formats).

Multigeneration: Copying material many times, often to create a multi-layered effect. In systems where successive generations causes loss of signal quality, such as those using analogue or digital picture compression, the use of multiple generations is approached with some caution.

Multi-image Lens: Multifaceted polished optical glass lens that produces multiple images of a subject in a single picture.

Multilanguage: A DVD-Video program containing soundtracks and subtitle tracks for more than one language.

Multimedia: A somewhat ambiguous term that describes the ability to combine audio, video and other information with graphics, control, storage and other features of computer-based systems. Applications include presentation, editing, interactive learning, games and conferencing.

Multi-pass Textures: The ability to "layer" textures onto the polygons, each with different tiling settings. For example, a brick texture tiled across a wall, and a single spray-paint graffiti texture on top, un-tiled.

Multi-path Reception: Interference to a primary radio or TV signal resulting from reflections from objects between the transmitter and receiver. Multi-path interference results in "ghosts" in TV reception.

Multiple Exposure: See Double Exposure.

Multiple System Operator: A company that operates more than one cable TV system.

Multiple-camera Production: A mode of production unique to television wherein two or more cameras are used to record the scene, enabling simultaneous and/or post-production editing. The mode used in most sitcoms and all soap operas, game shows, sports programs, and newscasts.

Multiple-image Shot: A shot that includes two or more separately photographed images within the frame.

Multiplex: A technique for putting two or more signals into a single channel.

Multiplexer: A mirror device which is part of a film island that selectively directs the light from multiple projectors into a single television camera.

Multi-programming: This refers to the capacity of a computer program to process more than one job at a time.

Multi-scan Monitor: A monitor (also referred to as multi-sync or multi-frequency) which synchronizes to different video signal sync frequencies, allowing its use with various computer video outputs.

Multi-screen Projection: Projecting motion picture images simultaneously on more than one screen.

Multi-session: A technique in write-once recording technology that allows additional data to be appended after data written in an earlier session.

Multi-track Recording: In the sound editing process, recordings are digitally or electronically divided into four (or many more) separate tracks. On each is a sound category (dialogue, music, effects) separated form the others, allowing the sound editor to manipulate individual soundtracks before producing a finished soundtrack.

Musco Lights: An array of permanently crane mounted HMI lights.

Music Editor: Individual who performs editing on the score of a movie.

Music Production Libraries: Collections of music

covering a wide variety of moods and needs, which have been especially designed for productions.

Music Supervisor: Individual who coordinates the work of the composer, the editor, and sound mixers.

Music Television: Generally refers to a system, such as a cable or satellite service (e.g., MTV, CMT), through which musical broadcast material is delivered.

Music Video: A visual representation of or accompaniment to a song or other musical selection that usually exists independently as a recording.

Musical Director: Person who selects and arranges the music for a program.

Musical Instrument Digital Interface: See MIDI.

Musical: A film genre that incorporates song and dance routines into the film story. Also called Musical Film.

Must-carry: Part of the 1992 Cable Act requiring cable systems to carry all local broadcast signals in their markets.

Mute Print: A positive print which carries the picture only (silent print).

Mute: Pictures without sound. Mute footage can be married with sound from another source such as a voice-over.

Mux: Short for multiplex.

Mux_rate: In MPEG, the combined rate of all packetized elementary streams (PES) of one program. The mux_rate of DVD is 10.08 Mbps.

Mythic Analysis: An interpretive strategy of genre analysis that approaches genres in terms of archetypes, stories shared by large segments of a culture which offer the researcher evidence of that society's thought process.

N

NA: See Numerical Aperture.

NAB: See National Association of Broadcasters.

NABET: See National Association of Broadcast Employees and Technicians.

Nanosecond: One-billionth of a second.

Narration (voice-over): When a character's or omniscient narrator's voice is heard over an image.

Narrative Enigma: A question that underpins a story and will (in classical films) or will not (in soap opera) be answered at the conclusion.

Narrative Function: A specific action or an attribute of a character in a narrative: according to the narrative theory of V.I. Propp.

Narrative Image: A particular representation of a program created by advertising and promotion in order to entice viewers.

Narrative Montage: Editing that constructs a story with film images by arranging shots in carefully sequenced order. See Montage.

Narrowcasting: Programming designed to reach a specific group defined by demographics and/or program content.

National Advertiser: An advertiser with national distribution whose products or services are on sale throughout most, if not all, of the country.

National Association of Broadcast Employees and Technicians (NABET): Union for technical personnel.

National Association of Broadcasters (NAB): The primary trade organization representing the interests of commercial broadcasting.

National Film Theatre: Main showcase cinema in the UK. Located in London, England.

National Rep: The link between the national advertiser, his agency, and the television station. The primary function of the rep is to create and develop sales of national advertising time for the television stations it represents.

National Television Standards Committee (NTSC): Created the first international television system for use in the U.S. and other countries which produces pictures by creating 525 alternating lines across the TV screen for each frame of video.

Native Format: For image capture and post production native format means the original sampling format of the material. For image display devices, native format means the display scanning format that all material is converted to prior to presentation.

Native Resolution: The resolution at which the video file was captured.

Naturalism: A style of filmmaking which is starkly realistic and which avoids any semblance of artifice.

Naturalistic Animation: An aesthetic tenet of animation which advocates that animation replicate live-action film/video as much as possible; cartoon characters should resemble objects in reality and our view of these characters/objects should resemble that of a camera. The opposite of abstract animation.

Naturalistic Performance: Performance style in which the actor attempts to create a character that the audience will accept as a plausible and believable human being, rather than an actor trying to portray someone.

Navigation Data: In DVD-Video there are five types of navigation data: Video Manager Information (VMGI), Video Title Set Information (VTSI), Program Chain Information (PGCI), Presentation Control Information (PCI) and Data Search Information (DSI).

Navigation Timer: In DVD-Video a system timer used during navigation operations.

NC-17 (No one Under 17 Admitted): A certificate issued by the MPAA indicating that no person aged 17 or under will be allowed to attend a screening of the movie. This category was formerly called "X," but many people's mistaken association of "X" films with XXX films caused the MPAA to change this on September 27, 1990.

NCTA: See National Cable Television Association.

ND Filter: An ND or Neutral density camera filter is used to reduce the intensity of light coming into the lens. A typical set of ND filters will reduce the exposure by 1, 2, 3-stops. The quality of light is not effected, only the intensity.

Needle Drop: The single use of a transcribed musical piece, generally for a fee.

Negative Cost: The cost of a movie through the production of a finished negative, not including the costs of prints, advertising, or distribution.

Negative Cutter: The person who cuts and assembles the original negative to match the edited workprint, which then goes to the lab for the answer print.

Negative Effect: Special effect in which either blacks and whites are reversed or colors are inverted. For example, red becomes a blue-green, green becomes purple, etc. An electronic color filter can be used for fine adjustment of the hues.

Negative Image: A photographic image in which dark and light tones are reversed, with dark areas appearing light on the screen and light areas appearing dark.

Negative Matcher: Same as Negative Cutter.

Negative Pickup: An agreement where a distributor acquires a finished negative and agrees to pay P&A and distribution costs. As opposed to deals where the distributor is involved during production.

Negative: The original film that is used in the camera, from which a positive print is made for editing. The negative is assembled to match the edited workprint, and an answer print, for projection of the completed film, is struck from the negative.

Negotiated Reading: In cultural studies, the interpretation of the text that partially accepts and partially rejects the meanings that the text emphasizes.

Neo-realism: An Italian film movement after World War II characterized by starkly realistic, humanistic stories and documentary-like camera style. Neo-realistic films were generally shot on location, using available lighting and non-professional actors. Also called Italian Neo-realism.

Nets: A bobbinet on a frame used to cut lighting intensity by either a half stop or full stop.

Network (Broadcasting): A group of television stations fed common programming from a central source. More specifically, the FCC defines a programming source as a "network" if it distributes at least 15 hours of programming a week to at least 25 affiliates in at least 10 states.

Network Advertising: The advertising placed by the network that the television station must agree to accept when they contract to carry the network programs. The network keeps most of the money and pays the television station in accordance with the size of their audience.

Network Compensation: The dollar amount the networks pay their affiliated television stations for broadcasting the network feed.

Network Feed: The sending of program material over network lines to affiliate television stations around the country. The feed may be aired live or taped and broadcast later.

Network Sale: The sale of a film to a particular television network to gain that network exclusive rights to the film for a certain period. Off-line editing which involves the transfer of tape to and from multiple sources. On-line Editing The final editing process in which the final video cut is merged with titles, sound effects and other special effects.

Neutral Density Filter: A filter which reduces the overall amount of light reaching the lens having no effect on color. Either used on the lens or/and on the light source.

Neutral Light: Planetary illumination from sun, moon, stars be it indoors or out. Has higher color temperature than artificial light, and thus more bluish qualities.

Neutral Set: A production set that does not suggest a specific setting or locale.

Neutral Shot: A head-on shot of a subject with no evident screen direction. By inserting a neutral shot between two successive shots taken from opposite sides of the line of action, screen direction can be changed.

New Wave: A group of French filmmakers in the late 1950's, the best known of whom were Francois Truffaut, Jean-Luc Goddard, and Alain Resnais. Also called Nouvelle Vague.

News Cooperative: A broadcast television station association in which news stories are exchanged between members.

News Director: The person responsible for the content, production and presentation of a newscast. Exact responsibilities depend on the news organization.

News Producer: Although responsibilities vary widely, producers are generally under the news director and in charge of specific newscasts. Producers often write basic transitional news copy.

Newsreel: A type of short film that presents a compilation of timely news stories.

Newsworthiness: Elements which make a news story interesting and noteworthy from the perspective of an audience.

NexTView: An electronic program guide (EPG) based on ETS 300707.

NHI: See Nielsen Television Station Index.

NiCad (Nickel Cadmium): Abbreviation coined and popularized by SAFT America for lightweight camcorder battery type designed to maintain power longer than traditional lead-acid batteries.

NICAM 728: A technique of implementing digital stereo audio for PAL video using another audio subcarrier. The bit rate is 728 kbps. NICAM 728 is also used to transmit non-audio digital data in China.

Nielsen Home Video Index (NHI): A Nielsen Media Research division primarily servicing the cable industry (basic and pay) as well as local cable operators. At present, NHI measures 29 basic cable networks and for pay cable networks. NHI uses data accumulated from the national People Meter sample.

Nielsen Syndication Service (NSS): A Nielsen Media Research division primarily servicing producers, national syndicators, as well as unwired networks, vignettes, and program producers. NSS uses data accumulated from the national People Meter sample.

Nielsen Television Index (NTI): A Nielsen Media Research division primarily servicing the four broadcast networks, advertisers, and advertising agencies. NTI uses data accumulated from the national People Meter sample.

Nielsen Television Station Index: (NSI): A Nielsen Media Research division primarily servicing local television stations, advertisers, advertising agencies, syndicators and producers with data from diaries or meter/diary samples. NSI conducts four "sweeps" in a year in which all local markets are measured as well as three mini-sweeps for the major markets (October, January and March).

Nielsen: The marketing research company that provides ratings for the television industry.

Night Viewing Device: An optical light amplifier that makes possible the taping of scenes under virtually no light.

Nighttime Mode: Name for Dolby Digital dynamic range compression feature to allow low-volume nighttime listening without losing legibility of dialog.

NLE: See Nonlinear Editing.

No Fax: A rehearsal without cameras, sets or lights.

Noddies: Cutaway shots of an interviewer's visual responses (nodding, smiling, etc.). Noddies are useful as transition devices during editing.

Noir: Usually refers to the classic black and white film noir style used in detective mysteries, typically employing hard lighting and dark, low key lighting.

Noise Figure: A measure of the performance (noise contribution) of an LNB in decibels. The lower the better.

Noise Floor: The level of background noise in a signal or the level of noise introduced by equipment or storage media below which the signal can't be isolated from the noise.

Noise Gate: A device used to modify a signal's noise characteristics. In video, noise gates provide optimal automatic suppression of snow (signal noise level). In audio, a noise gate provides a setable signal level threshold below which all sound is removed.

Noise Reduction: An electronic process used to reduce noise levels in audio and video. In video, the most effective noise reduction is accomplished by digitizing the video signal and carrying out a computerized pixel by pixel analysis of the data. In audio, the most effective systems employ an encode/decode scheme, performed before and after recording, such as the Dolby audio noise reduction system. Noise reduction can be performed on an existing audio signal using systems such as DNR (dynamic noise reduction) but are less effective because they also affect the audio signal.

Noise: A general term used in electronics to indicate any unwanted electrical signal, unrelated to the original signal. Video noise is generally manifested as snow, graininess, ghost images or picture static induced by external sources such as the national power-line grid, electric motors, fluorescent lamps, etc. In audio, noise is generally manifested as hiss and static.

Nominal Viewing Distance: Distance between viewer and TV picture at which the TV raster line structure becomes invisible. For 525 line systems, the recommended viewing distance is from 5 to 7 times screen height.

Non-additive Mix: A mix of two pictures which is controlled by their luminance levels relative to each other as well as a set mix percentage (or position of the lever arm). A and B sources are scaled by factors K and 1-K but the output signal is switched to that which has the greatest instantaneous product of the scaling and the luminance values. The output of any pixel is either signal A or B but not a mix of each. So if K = 0.5, where picture A is brighter than B then only A will be seen. Thus two clips of single subjects shot against a black background can be placed in one picture. The term has come to encompass some of the other more exotic types of picture mixing that are available today.

Non-composite Video Signal: Video signal containing picture and blanking information but no sync signals.

Non-diegetic Sound: Sound that does not occur in the diegetic space (the characters' world), such as music that is added in post-production.

Non-directional Microphone: A microphone with a circular polar pattern of response, equally sensitive to sound from all directions.

Non-diversity Antenna: Use of a single receiving antenna for a RF microphone system. See also Diversity System.

Non-Drop Frame: Type of SMPTE time code that continuously counts a full 30 frames per second.

Nonfiction Film: Any film that does not employ invented plot or characters. See Documentary.

Noninterlace: This is a method of scanning out a video display that is the total opposite of interlaced. All of the lines in the frame are scanned out sequentially, one right after the other. The term "field" does not apply in a non-interlaced system. Another term for a non-interlaced system is progressive scan.

Non-interlaced Video: Process of scanning complete frames in one pass, painting every line on the screen, yielding higher picture quality than that of interlaced video. Most computers produce a non-interlaced video signal; NTSC is interlaced.

Nonlinear Editing (NLE): Editing performed on a computer, in which shots do not have to be placed one after the other (i.e., in a linear fashion).

Nonlinear Editor: An editing system based on storage of video and audio on computer disk, where the order or lengths of scenes can be changed without the necessity of re-assembling or copying the program.

Non-narrative Television: Televisual texts (e.g., news and sports programs, game shows, some commercials) that present reality to us without using conventional narrative structures.

Non-Reflex: A camera that does not have a "through the lens" viewfinding system, but gives you an image in the viewfinder through a separate lens. Older Bolexes and Bell & Howell cameras are non-reflex.

Non-segmented: A video recording process in which a complete (as opposed to a partial) video frame is recorded or played back with each pass of a video head.

Non-synchronous Sound: Sound whose source is not apparent in a film scene or which is detached from its source in the scene; commonly called off-screen sound.

Non-WGA: Written work for the entertainment industry

that is done for a company which has not signed (or become a signatory} to the Writers Guild of America Minimum Basic Agreement.

Normal Lens: A medium-focal length lens which minimizes perspective distortion. In 16mm this is the 25mm lens. In 35mm it is the 50mm lens. It is the point between the widening of the image by the wide angle lens and the magnifying of the image by the telephoto lens.

Nose Grease: Just what it sounds like. Used in the old trick among camerapersons to lubricate the pressure plate by wiping it along the side of the nose.

Nose Room: The space between the nose of a subject's head in profile and the edge of the video frame the subject is facing. The general rule is that the space around the subject should be apportioned to 2/3rds in front of the subject's head, and 1/3rd behind.

Notch: A recess on the edge of a piece of film which automatically triggers a mechanism effecting some modification of the duplication process, commonly a change of exposure light intensity.

NRZI (Non-Return to Zero, Inverted): A method of coding binary data as waveform pulses. Each transition represents a one, while lack of a transition represents a run of zeros.

NSI: See Nielsen Television Station Index.

NSS: See Nielsen Syndication Service.

NTI: See Nielsen Television Index.

NTSC 4.43: This is a NTSC video signal that uses the PAL color subcarrier frequency (about 4.43 MHz). It was developed by Sony in the 1970s to more easily adapt European receivers to accept NTSC signals.

NTSC Standard: Normally refers to the 525-line, 60-field system of broadcast television which combines chroma with luminance information into one composite signal.

NTSC: See National Television Standards Committee.

NTSC-4.43: A variation of NTSC where a 525/59.94 signal is encoded using the PAL subcarrier frequency and chroma modulation. Also called 60-Hz PAL.

Numerical Aperture (NA): A unit-less measure of the ability of a lens to gather and focus light.

Nut: Operating expenses to be recovered, often deductible in an exhibitor's contract with distributors.

Nuvicon: A type of video pickup tube used in inexpensive color television cameras.

Nyquist Frequency: The Nyquist rate is considered the minimum sampling frequency for correct digital reproduction of a given signal or twice the signal bandwidth. Alternatively, the Nyquist frequency is the highest signal frequency which can be correctly reproduced for a given sampling rate.

Nyquist Sampling Theorem: Intervals between successive samples must be equal to or less than one-half the period of the highest frequency.

O

O&O: A television station owned and operated by a national network.

Obie: An eyelight mounted on the camera.

Objective Camera: A sustained camera angle that approximates the viewpoint of a human observer.

Objective Correlative: An object which comes to represent an aspect of a character; e.g., Bart Simpson's skateboard representing his carefree and spontaneous lifestyle.

Oblique Angle: A shot taken from a tilted camera which, when projected on the screen, looks as if it is titled on a diagonal.

Observational Mode: Type of television text wherein a television producer's presence is not obvious to the viewer, and his or her manipulation of the historical world is minimal.

OC-3: An OC-3 circuit (155,000,000 bits per second) is the nominal backbone speed that major national ISPs use in the United States.

OC-48: An OC-48 circuit (2,400,000,000 bits or 2.4 gigabits per-second) is the typical speed for many aggregated telephone voice circuits on inter city fiber optic lines.

OCN: See Original Camera Negative.

Octave: The interval between two sounds having a basic frequency ratio of 2 to 1.

Off Network Syndication: Commonly known as "re-runs," programs that have aired at least once on a network and are now available for any television station to purchase the rights to air it.

Off Screen (OS): Dialogue or sounds heard while the **Off-**

Line Editor: A low resolution, usually computer and disk based edit system in which the creative editing decisions can be made at lower cost and often with greater flexibility than in an expensive fully equipped on-line bay.

Off-Scale: Outside the range of the standard light values of a printer.

Off-screen Sound: See Non-synchronous Sound.

Off-screen Space: Space beyond the camera's field of vision which the audience is aware of nevertheless.

Offset: Type of dish with the focus and feedhorn below the center of the dish.

OHM: A measure of electrical resistance or impedance.

OIRT: Organisation Internationale de Radiodiffision-Television.

Omega Wrap: The tape path used in some helical videotape machines that resembles Greek letter Omega.

Omni-directional Microphone: A microphone that is able to pick up sound equally from all directions.

Omniscient Point of View: The all-knowing narrator.

On-air Director: The technical director. Primarily responsible for video switching during a production.

On-Demand Streaming: Sending video/audio or other data that is transmitted to the end-user upon request. Widely used on news and sports-oriented Web sites, for example.

One Light Print: The alternative to a Timed Print, a one light is a print that has not been corrected shot by shot, but shows what all the shots look like with the same printing lights in contrast to each other.

One_Random_PGC Title: In DVD-Video, a Title within a Video Title Set (VTS) that contains a single Program Chain (PGC), but does not meet the requirements of a One_Sequential_PGC Title.

One_Sequential_PGC Title: In DVD-Video, a Title within a Video Title Set (VTS) that contains a single Program Chain (PGC) with the following attributes: 1) PG Playback mode is Sequential, 2) no Next PGC, Previous PGC or Go Up PGCs are defined, and 3) the Navigation Timer is neither set, nor referred to.

On-line (editing): Production of the final edit performed at full program quality. Being higher quality than off-line, time costs more. Preparation in an off-line suite will help save time and money in the on-line.

On-Line Editor: An editing system where the actual video master is created. An on-line bay usually consists of an editing computer, video switcher, audio mixer, 1 or more channels of DVE, character generator, and several videotape machines.

Opacity: The ratio of the amount of light falling on a surface to the amount of light which is transmitted.

Opaque Leader: Any strip of flexible, optically opaque material used to space picture in A/B roll film cutting and editing.

Open Architecture: A computer design that allows for the easy addition of circuit boards produced by a variety of manufactures.

Open Forms: A frame in which part of the action is cut off or masked.

Open Subtitles: See Subtitles.

Open: The time at which a movie is first released in theatres.

Opening Weekend: When a movie is first released in each country, its success is often measured in terms of its gross for the first weekend it opened. A disproportionate number of people usually see a movie on its opening weekend so box-office numbers are a good guide as to whether the movie will be a hit or not.

Opposite Track Path (OTP): Dual-layer disc where Layer 0 and Layer 1 have opposite track directions. Layer 0 reads from the inside to the outside of the disc, whereas Layer 1 reads from the outside to the inside. The disc always spins clockwise, regardless of track structure or layers. This mode facilitates movie playback by allowing seamless (or near-seamless) transition from one layer to another.

Oppositional Reading: In cultural studies, the interpretation of the text that is wholly contrary to the text's dominant meanings.

Optical Disc: A recording medium for audio and/or video where a laser beam is used to record and play back digitally-coded information.

Optical Effects: A laboratory or print procedure in which shots are modified by use of an optical printer. These are most commonly seen as fades and dissolves; however, it can include a wide range of special effects procedures.

Optical Fiber: A glass strand conductor designed to conduct light waves modulated with audio or video information. See also Fiber Optic.

Optical Printer: Elaborate film laboratory equipment, consisting of a projector and a camera, which can reduce or enlarge film images and create special effects—such as fades, dissolves, and superimposition—in a printed film.

Optical Sound: The system used by a projector to play back sound from a film print. The sound is exposed onto the film as a clear modulating line against black. It corresponds to the modulations of the sound. The projector reads the track by passing it between the exciter lamp and light-sensitive photo-electric cell which generates a voltage that is amplified and fed into a speaker.

Optical Soundtrack: A composite print in which the soundtrack is recorded via the varying width of a transparent track which runs beside the sequence of frames on a print. See also Advance.

Optical Track: An intermediate step from going from the mix master to the final print is to have an optical track struck. An optical track is photographed onto a blank piece of special high contract stock by the facility where the mix is done, or by the lab. The optical track is a separate roll of film from the original negative and is combined with picture when a print is struck.

Optical Viewfinder: A camera viewfinder which has no electronics and therefore cannot reflect the actual picture being transmitted from the camera. Optical viewfinders for inexpensive television cameras are rangefinder viewfinders which are separate from the camera lens system and show a picture slightly different from that seen by the camera. The difference is most troublesome for subjects close to the camera where the angle of difference, or parallax, is greatest.

Optical: Any device carried out by the optical department of a lab using an optical printer. (i.e.,. dissolves, wipes, double exposure effects).

Option: Money paid to a writer by a producer for exclusive rights to a script. A specific time is specified during which the producer tries to put together funding to put the project into production.

Orange Book: The document begun in 1990 which specifies the format of recordable CD. Three parts define magneto-optical erasable (MO) and write-once (WO), dye-sublimation write-once (CD-R), and phase-change rewritable (CD-RW) discs. Orange Book added multi-session capabilities to the CD-ROM XA format.

Orbit: A plan that guarantees the buyer that a spot will run in every program included in the orbit. The number of spots is always equal to the number of programs included in the orbit, with each program given equal weight when computing ratings and demographics (as opposed to a rotation that gives equal weight to each available break).

Order Clearance: The determination made by either the sales or traffic departments as to whether an advertiser's schedule can be fit into the existing avails. If the order can be placed without causing an oversell, or if the television station allows the schedule to displace a schedule of lesser value, the order is said to have "cleared."

Original Camera Negative (OCN): The negative film originally passed through the camera.

Original Color Negative (OCN): A developed negative.

Original: Any film, negative or reversal that was shot by a camera, as opposed to a print or intermediate copy. The term original can be used interchangeably with negative, but is as especially handy term when taking about reversal film, where it is the clearest way indicating whether something is a dupe or the original.

Orthochromatic Stock: A kind of black and white film stock that reacts particularly to the blue and green areas of the color spectrum rather than the red; widely replaced by panchromatic stock after 1926, but still used for special applications.

Orthogonal Sampling: A sampling method done in phase with the horizontal line such that sample positions are aligned vertically in the image.

OS: See Off Screen.

Oscars (Academy of Motion Picture Arts and Sciences Awards): The term "Oscar" was coined by an anonymous person who remarked that the statue looked like their Uncle Oscar.

OS-shot: A shot taken over the shoulder of one person, focusing on the face of a second person.

OSTA: Optical Storage Technology Association.

OTO: The abbreviation for one-time-only, this refers to spots or programs that are to air only once.

OTP: See Opposite Track Path.

Out of Band: In a place not normally accessible.

Outdated Stock: Film is perishable. When it starts getting stale the dyes will shift color and the grain will build up, giving you a generally fogged, muddy and desaturated effect. It is only after about 2 or 3 years that this will start to happen, provided the film is refrigerated. Faster films tend to become outdated slightly faster than slow films. Likewise, color film will become outdated a little sooner than black and white.

Outer Diameter: Width of the disc. This is 12 cm for "normal" CDs and DVDs, and 8 cm for small CDs and DVDs.

Outline: An expanded synopsis of the story created by the writer. Usually just a few pages in length, it details the main plot, sub-plots and the relationships between the main characters.

Out-point: The last frame of footage required for inclusion in an edit.

Output: Processed data in either displayed or printed form, or stored on disk or diskette.

Out-take: Any footage deleted from a film during editing; more specifically, a shot or scene that is removed from a film before the final cut. Very small bits, a few frames or as little as one frame, are known as Trims.

Over Record: Recording more footage than is required before the in-point and/or after the out-point. This can be very useful in a true random access suite as the excess gives room for later adjustments. In a VTR/linear suite over recording could mean the program would have to be re-edited, a time-consuming operation.

Over the Shoulder Shot: View of primary camera subject framed by another subject's shoulder and back of head in foreground. Common in interview situations, perspective affords sense of depth.

Overcrank: To run the camera faster, producing slow motion. The term has survived from the time when you would crank a camera.

Overexposure: Filming a scene with more light than the

emulsion of the film can easily tolerate. The image will be too light and there will be less depth of field than if the lens had been set correctly. If compensated for in printing, the image will appear contrasty.

Overhead Expenses: Basic on-going costs related to maintaining a production facility, not directly related to a specific production.

Overhead Shot: A camera shot from directly above the action. See Bird's-eye View.

Overlapping Editing: Cutting that repeats part or all of an action.

Overlay: Keyed insertion of one image into another. Overlay is used for example, to superimpose computer generated text on a video image, for titling purposes. In video, the overlay procedure requires synchronized sources for proper operation.

Overnights: Nielsen Media Research's service that provides next-day feedback on television viewing in most of the 40 largest markets.

Oversampling: Use of a higher than necessary sampling rate in converters to simplify analog filtering.

Overscan: A loss of some information around the outer edges of a picture, typically, as a result of the broadcast and reception process.

Overtone: Acoustical or electrical frequencies that are multiples of the fundamental frequency or sound. Overtones add richness and uniqueness to sound.

Oxide Tape: Magnetic tape coated with fine particles of manganese oxide.

Oxide: Magnetic coating on video and audio tapes that stores picture and sound information.

P

P&A: Prints and advertising, the major costs of film distribution.

PA System: Public address (loudspeaker) system. Often used as part of a studio talkback system.

PA: See Production Assistant.

Pack: A set of clips, mattes and settings for DVE, color corrector, keyer, etc. that are used in processing. Quantel equipment allows packs to be saved and archived so they can be used later for re-works.

Package News Package: A completed news segment containing all needed audio and video components, ready for insertion (rolling into) a news production.

Package Plan: An advertising plan that bills the advertiser a flat rate for an entire spot schedule instead of a unit rate per spot (e.g., a six month schedule for 20 spots per month at $1000.00 a month = $6,000.00). A package plan may be designed for any or all lines on a contract. The system bills each line according to the package information that is entered with the contract.

Package: In television journalism, an 80-105 second news story shot in the field and filed by a reporter.

Packaging: Process of putting together on the same project a writer or screenplay, director, producer, and star talent, all of whom are generally represented by the same agency, and presenting this package to a studio.

Packet: A low-level unit of DVD-Video (MPEG) data storage containing contiguous bytes of data belonging to a single elementary stream such as video, audio, control, and so forth. Packets are grouped into packs.

Packetized Elementary Stream (PES): The low-level stream of MPEG packets containing an elementary stream, such as audio or video.

Packetized Elementary Stream: See PES.

Packing the Tape: Completely fast-forwarding a videotape to the end and then rewinding it. It is recommended that this is done to equalize tension, dislodge loose tape oxide and reduce recording problems before videotapes are used for the first time.

Paint Pots: Controls that adjust video color.

PAL (Phase Alternate Line): The European color TV broadcasting standard featuring 625 lines per frame and 25 frames per second. See NTSC and SECAM.

PAL 60: This is an NTSC signal that uses the PAL color subcarrier frequency (about 4.43 MHz) and PAL-type color modulation. It is a further adaptation of NTSC 4.43, modifying the color modulation in addition to changing the color subcarrier frequency. It was developed by JVC in the 1980s for the use with their video disc players, hence the early name of "DISK –PAL." There is a little-used variation, also called PAL 60, which is a PAL video signal that uses the NTSC color subcarrier frequency (about 3.58 MHz) , and PAL-type color modulation.

Palette: A table of colors that identifies a subset from a larger range of colors. The small number of colors in the palette allows fewer bits to be used to represent each pixel. Also called a Color Look-up Table (CLUT).

PALplus: A widescreen (16x9) television standard in use in Europe that is compatible with existing 4x3 TV sets. Non-16x9 TVs show the picture in a letterboxed form.

Pan and Scan: The technique of chopping off strips from one or both sides of a picture so it will be properly displayed on a TV screen. As the aspect ratio of movies are rarely the same as the aspect ratio of a television screen, when showing movies on TV it is necessary to make sacrifices. "Pan and Scan" refers to the technique of chopping off strips from one or both sides of the picture when displaying. The areas chopped off are typically changed on a shot-by-shot basis, depending on scene composition. The main advantage of this technique is that it allows detail to be seen, the disadvantage is that shot composition is sometimes destroyed.

Pan Handle: The handle attached to the pan head of a camera that enables the cameraperson to pan or tilt the camera.

Pan Pot: An attenuator-based device which can "place" a sound to varying degrees in a left or right stereo channel by raising or lowering the volume of the sound.

Pan Stick: A grease-based foundation makeup commonly used to cover a beard or skin blemishes.

Pan: Short for "panorama." A shot where the camera pivots horizontally, turning from left to right or from right to left. Also called Panning Shot.

Panaglide: A brand-name for a body camera mount which uses a system of counterbalanced springs to keep a camera reasonably steady, even when the camera operator is walking or running.

Panasonic 5-pin edit Control: Panasonic's wired edit control protocol. See Control-M.

Pancake: A type of makeup base. Pancake makeup is usually water based and applied with a sponge.

Panchromatic Stock: Black/white film stock that is sensitive to all the colors of the spectrum, from red to blue, but less capable of achieving great depth of field than the orthochromatic and monochromatic stocks it replaced.

Pantograph: Expandable device for hanging lighting instruments and TV monitors from a studio lighting grid.

Pantomime: A style of naturalist performance in which the actor presents the character with specific gestures that, through convention, represent specific emotions or actions.

Paper and Pencil Editing: Logging time codes by hand on an editing log as opposed to entering the decisions directly in a videotape editing system.

Paper Tape: A skinny roll of tape used to tape down the ends of film when editing, called paper tape to distinguish it from splicing tape.

Parabola: A parabolic dish used to reflect sound waves, concentrating them on a microphone, allowing sound to be picked up from greater distances than with even a normal unidirectional or shotgun microphone.

Parabolic Microphone: A highly directional microphone assembly that uses a reflector in the shape of a parabola to focus sound into a microphone.

Paradigmatic Structure: In semiotics, a manner in which signs are organized and meaning created. Paradigmatic structures create meaning through association, in contrast to syntagmatic structures which create meaning through sequence or chronological order. E.g., in baseball, the players that might replace one another in the batting line-up are in paradigmatic relationship to one another. The television viewer flows syntagmatically through the evening's programs (one after another) and surfs paradigmatically across the channels (hitting that remote control) to view what is on concurrently.

Parallax: The difference in view caused by looking at a scene from two slightly different locations.

Parallel Action: A device of narrative construction in which the development of two pieces of action are presented simultaneously. See also Cross-cutting.

Parallel Cutting: The alternate intercutting of two related stories so that each is regularly updated and both reach a conclusion at about the same time. The stories are generally related in some way.

Parallel Editing: The technique of intercutting between two simultaneous stories or scenes.

Parallel Stories: See Parallel Cutting.

Parallel Track Path (PTP): A variation of DVD dual-layer disc layout where readout begins at the center of the disc for both layers. Designed for separate programs (such as a widescreen and a pan & scan version on the same disc side) or programs with a variation on the second layer. Also most efficient for DVD-ROM random-access application.

Parallels: Temporary Scaffolding, used as a platform for the camera, lighting, or other rigging.

Params: An abbreviation for the parameters that are used to adapt software to the unique requirements of a television station.

Parental Management: An optional feature of DVD-Video that prohibits programs from being viewed or substitutes different scenes within a program depending on the parental level set in the player. Parental control requires that parental levels and additional material (if necessary) be encoded on the disc.

Parity: A simple form of data error detection based on adding an extra bit to each digital word. This extra bit is derived by a counting of the zeros or ones in a word. Data losses during transmission can thereby be detected.

Part of Title (PTT): In DVD-Video, a division of a Title representing a scene. Also called a chapter. Parts of titles are numbered 1 to 99 in a One_Sequential_PGC Title and 1 to 999 in a Multi_PGC Title.

Participating Dealers: The term used to describe the purchase of an advertising schedule by several non-competitive advertisers who share the charges and receive individual invoices.

Particles: Pixels drawn in a 3-D system to create special effects, such as rain, snow, explosions, etc. In increased density, can also be used for effects like smoke, fire, and waterfalls.

Patch Cord: The cord on a patch panel used to complete a connection between various audio or video sources and desired pieces of equipment.

Patch Panel: A control panel where all the video and audio lines used in a studio are brought together and terminated in connectors allowing any combination of lines to be wired together as desired by patching in short lengths of cable.

Patch: Using cables or electronic circuits to interconnect video and audio equipment.

Path: The trail along which a computer-graphics object travels.

Pattern Change: A notation in the log at dusk and dawn for AM television station engineers to change the transmission pattern so their signal does not interfere with other television stations during the night.

Pattern Projector: An ellipsoidal-type spotlight which includes a cookie (cucalorus) insert. Used to create a wide variety of designs on a background.

Pattern: A pair of cardioid microphones or elements aimed in crossed directions which feed two channels for stereo pickup.

Pay or Play: A contract provision which commits the production company to compensate a cast or crew member for a project whether or not that project ever goes into production.

Pay-Per-View (PPV): Programs purchased by a cable subscriber on a per-program basis rather than per-month basis.

Pb, Pr: Refer to the color component video signals B-Y and R-Y respectively with optimization for analog component video purposes or transmission.

PBS: See Public Broadcasting Service.

PCI (Presentation Control Information): A DVD-Video data stream containing details of the timing and presentation of a program (aspect ratio, angle change, menu highlight and selection information, and so on). PCI and DSI together make up an overhead of about 1 Mbps.

PCM: See Pulse Code Modulation.

PC-TV: The merger of television and computers. A personal computer capable of displaying video as a television.

PDC: See Program Delivery Control.

Peak Program Level Meter (PPM): A meter that monitors audio levels and, unlike a normal VU meter, is able to accurately respond to rapid transient peaks (changes) in volume.

Pedding: Vertical camera movement, rising or lowering, with camera levelness maintained. Approaching closer to either floor or ceiling, the up/down equivalent of dollying. Sometimes referred to as Ped Up and Ped Down.

Pedestal Level: A black level which is the absolute black level of a video signal. Normally, a video signal refers to the setup level (about 0% to 5% of video amplitude above the blanking level) as the black level.

Pedestal: A camera support generally restricted to studio use having a single elevator column mounted on a tricycle base.

Pel: See Pixel.

Pels (Picture elements): The smallest dots that can be transmitted for any given bandwidth and scanning system.

Penetration: In a given population, the percent of households using a product or receiving a service. Most commonly used to define the number of households that are subscribing to cable.

People Meter: A device attached to a television that records which channel is being viewed by that household. There are typically several buttons on the box, one for each family member, which they press each time they enter or leave the room. It provides the ratings company with demographic information as well as the channel tuning data.

Perambulator Boom: A large microphone boom

extending 30 or so feet, normally mounted on a dolly, that permits rapid and quiet relocation of a microphone anywhere on a set.

Perceived Resolution: The apparent resolution of a display from the observer's point of view, based on viewing distance, viewing conditions, and physical resolution of the display.

Perceptual Coding: Lossy compression techniques based on the study of human perception. Perceptual coding systems identify and remove information that is least likely to be missed by the average human observer.

Perf: Perforations. The sprocket holes in a piece of film.

Perfect Fit: In the study of television stars, a matching of a particular role's characteristics to a star's polysemy.

Performance Continuity: See Continuity.

Performer: On-camera talent. A person who appears on camera in non-dramatic productions.

Performing Rights License: General permission, normally by ASCAP (American Society of Composers, Authors, and Publishers) and BMI (Broadcast Music Incorporated), to play copyrighted music. Such licenses only cover general music-centered programming and not commercials, public service announcements, and productions.

Periaktos: A triangular piece of scenery.

Per-Inquiry Advertising (PI): Advertising for which the advertiser is not billed at the card rate, but is charged a certain percentage for all money received on mail order sales of the advertising product. Examples of PI advertiser are Pocket Fisherman, Veg-O-Matics, Ginsu Knives, etc.

Peripherals: Accessories and auxiliary computer attachments that expand a computer's capabilities.

Persistence of Vision: Tendency of human vision to retain images for a fraction of a second. Discrete images presented at a rate of about 16 or more per second—even when change takes place between them—blend together, creating the illusion of motion in TV and film.

Personal Microphone: A small, individual microphone, generally attached to clothing with a clip.

Personality Star: An actor who plays only those roles associated with his or her public image. For example, John Wayne, Marilyn Monroe, and Barbara Streisand are personality stars.

Persons Using Television (PUT): PUT is the percentage of a target audience tuned in to any programming during a specific time period.

Perspective: Visual phenomena in which parallel lines appear to coverage as camera-to-subject distance increases.

PES (Packetized Elementary Stream): A coded MPEG bitstream for one program of audio or video or one data stream. PES includes stamps and header information.

PES: See Packetized Elementary Stream.

P-Frame: In a video compression scheme, the difference between the complete subject frame and its predecessor. This is the most compact amount of data needed to generate the subject frame given the complete frame preceding, but it does not allow backwards single step or playback. See also I-Frame.

PG: A certificate issued by the MPAA indicating that a film's content is suitable for viewing by children, but recommends parental guidance. See also PG-13.

PG-13: A certificate introduced by the MPAA on July 1, 1984 to indicate that a movie's content is rated as slightly stronger than a PG certificate. See also R.

PGC: See Program Chain.

Phantom Power Supply: Power which comes from a mixer that is used to operate a condenser microphone.

Phase Adjust: This is a term used to describe a method of adjusting the hue in a NTSC video signal. The phase of the color subcarrier is moved, or adjusted, relative to the colorburst. PAL and SECAM systems do not usually have a phase (or hue) adjust control.

Phase Cancellation: Interference resulting in reduced audio levels caused by a sound source being picked by two microphones.

Phase Distortion: This is a shifting of output voltage relative to input by an amount which is disproportional to frequency. This will not detectable until it an amplifier.

Phase Error: A change in the color subcarrier signal which moves its timing out of phase, i.e., it occurs at a different instant from the original signal. Since color information is encoded in a video signal as a relation between the color subcarrier and the colorburst phase, a deviation in the color subcarrier phase results in a change in the image's hue.

Phase Shift: The displacement of a waveform in time. Some electrical components introduce phase shift into a signal. When various frequencies are displaced differently, distortion occurs. Electrical cancellation may occur when two equal signals are out of phase by 1~3()¡. However, this may also be used are a encoding method where the shift is removed on playback (similar to the method in which a Dolby Surround soundtrack is encoded and played back.

Phase: Timing relationship between two signals.

Phase-change: A technology for rewritable optical discs using a physical effect in which a laser beam heats a recording material to reversibly change an area from an amorphous state to a crystalline state, or vice versa. Continuous heat just above the melting point creates the crystalline state (an erasure), while high heat followed by rapid cooling creates the amorphous state (a mark). (Other recording technologies include dye-sublimation and magneto-optical.)

Phi Phenomenon: A physiological effect that, along with persistence of vision, creates the illusion of motion when consecutive images are presented in a rapid sequence.

Phone Plug: Sturdy male connector compatible with audio accessories, particularly for insertion of microphone and headphone cables. Not to be confused with phono plug.

Phono Plug: Also called "RCA" or "RCA phono," popular cable connector for home audio as well as video components. Standard connection for direct audio/video inputs/outputs. Not to be confused with phone plug.

Phosphor Aging: Reduction in the amount of light emitted by phosphors due to deterioration with use.

Phosphor: Material that glows within a CRT when struck by the electron beam phosphor composition determines the maximum range of colors that can be displayed on a screen.

Photo CD: Kodak's Photo CD for representing 24-bit 4:2:0 YCbCr images hierarchically at resolutions of up to 3072x2048 pixels. Thumbnails image representation is also part of the Photo CD spec. Built upon CD-ROM XA.

Photo Flood: A photo flood is a high power screw-in light bulb that is often used in with a clamp light fixture. Photo floods are usually anywhere from 250 watts to 500 watts.

Photoemission: The process of converting light—generally as focused on a camera target—into electrical energy.

Photoflood Lamp: A lamp that resembles a normal light bulb except that it produces a much higher light output. Photoflood lamps normally last only about six hours.

Photographic Lens Tissue: This form of tissue is used in conjunction with a lens cleaning fluid. The tissue is very soft so that the risk scratches to the lens can be minimized.

Photo-realistic: A live action image would be considered photo-realistic. Usually the term is reserved to the creation of a computer generated image (CGI) that is created with the details and quality that it would appear to be an actual photo.

Physical Format: The low-level characteristics of the

DVD-ROM and DVD-Video standards, including pits on the disc, location of data, and organization of data according to physical position.

Physical Sector Number: Serial number assigned to physical sectors on a DVD disc. Serial incremented numbers are assigned to sectors from the head sector in the Data Area as 30000h from the start of the Lead In Area to the end of the Lead Out Area.

Pickup Pattern: In microphones, the shape of the space in which the microphone is sensitive to sound. Common patterns include omni-directional and cardioid.

Pickup Shot: Reshooting a portion of a scene, the rest of which was acceptably filmed in a previous take.

Pickup Tube: A light-sensitive electron tube which is scanned by an electron beam to convert an image focused on the face of the tube into an electronic signal.

Pickup: A process of matching time codes or pictures to enable a match-frame edit. Also movies made by one studio that have been acquired by another.

Picture Height: Reference distance measurement for resolution of TV screens, used for horizontal as well as vertical discussion. So chosen because resolution is correctly measured in the largest circle that fits in the area referred to and for a TV screen the diameter of such a circle equals "one picture height."

Picture Sharpness: The fine details in a video picture. A picture appears sharp when it contains fine details and has good contrast. Picture sharpness is easily lost during the recording/playback process. Advanced video enhancement equipment is used to improve picture sharpness, especially contrast, and can pre-compensate for potential losses which might alter an image during processing.

Picture Stop: A function of DVD-Video where a code indicates that video playback should stop and a still picture be displayed.

Picture: In video terms, a single still image or a sequence of moving images. Picture generally refers to a frame, but for interlaced frames may refer instead to a field of the frame. In a more general sense, picture refers to the entire image shown on a video display.

Picture: The workprint, to distinguish it from the mag tracks.

Pigeon: This is a heavy round disc with a lighting stud, used to position a light on the floor, much lower than a stand will go. Basically, it is a Hi Hat for lights.

Piggyback: A commercial announcement, half of which promotes one product or service, and the other half promotes another product or service, but both for the same company (i.e., a 90 second spot containing two 45 second commercial announcements back-to-back).

Pilot Signal: A signal from a satellite transponder used by ground television stations to align up-link dishes.

Pilot Tone: A sine wave signal, recorded by various field audio recorders at a known frequency, which is used to resolve the tape speed on playback to retain sync with film camera footage.

Pilot: A program, sometimes a made-for-TV movie, which introduces a new series.

Pin: A component of a camera or printer mechanism which engages with a perforation hole to move and locate film for exposure.

Pink Noise: A sound signal that has an equal amount of energy per octave or fraction of an octave.

PIP (Picture in a Picture): A feature on higher priced TV sets whereby a small picture for a second program can be displayed in one corner of the screen. The viewer can switch back and forth as to which program occupies the whole screen. In order for two broadcasts to be so seen simultaneously the TV set must have two tuners (channel selectors). More commonly there is only one tuner and the second program must come from a VCR or other local source.

Pipeline: A schedule of movie projects in production.

Pirate: To illegally copy or broadcast a copyrighted production without permission.

Pit Art: A pattern of pits to be stamped onto a disc to provide visual art rather than data. A cheaper alternative to a printed label.

Pit Length: Arc length of pit along the direction of the track.

Pit: A microscopic depression in the recording layer of a disc. Pits are usually 1/4 of the laser wavelength so as to cause cancellation of the beam by diffraction.

Pitch: Meeting in which one party will attempt to interest another party in a particular work or in a version of a particular work by presenting the story of the work in such an exciting manner that the buying party will find great interest in the work and will either buy the work or pay the "pitching" party to write the work. Pitch is also the distance between two successive perforations along a strip of film. It is also a term used to describe the frequency of audible sound.

Pix: An abbreviation for picture used on the leader.

Pixel Aspect Ratio: The ratio of width to height of a single pixel. Often means sample pitch aspect ratio (when referring to sampled digital video). Pixel aspect ratio for a given raster can be calculated as y/x multiplied by w/h (where x and y are the raster horizontal pixel count and vertical pixel count, and w and h are the display aspect ratio width and height). Pixel aspect ratios are also confusingly calculated as x/y multiplied by w/h, giving a height-to-width ratio.

Pixel Clock: The pixel clock is used to divide the horizontal line of video into samples. The pixel clock has to be stable (a very small amount of jitter) relative to the video or the image will not be stored correctly. The higher the frequency of the pixel clock, the more samples per line there are.

Pixel Depth: Pixel Depth is the number of bits that each pixel can represent.

Pixel Drop Out: This can be a real troublemaker, since it can cause artifacts. In some instances, a pixel drop out looks like black spots on the screen, either television stationary or moving around. Several things can cause pixel drop out, such as the ADC not digitizing the video correctly. Also, the timing between the ADC and the frame buffer might not be correct, causing the wrong number to be stored in memory. For that matter, the timing anywhere in the video stream might cause a pixel drop out.

Pixel: A tiny sample of video information, the "dots" that make up an overall picture. The pixel is the smallest division or element that a single image will be divided. In our discussions, we assume square pixels. With square pixels, we can easily discuss the effects of pixel resolution on aspect ration.

Pixilation: A type of film animation in which real objects or people are photographed frame by frame in order to make them appear to move abruptly or magically when the film is projected. See also stop motion and trick film.

Plastic Leader: This is leader for putting at the head and tail of a print. It is, as one would guess, made out of plastic, and is more durable than emulsion leader and much less expensive, and so it is the better choice for a print. However, it cannot be cement spliced, so it should not be used for the negative.

Plate: A background for any type of process shot.

Playback Head: The device on a audio or videotape machine responsible for detecting prerecorded material.

Playback: A technique of filming music action first, the playing the music through loudspeakers while performers dance, sing, etc.

Play-by-play Announcer: A type of television sports announcer, usually a professional broadcaster, who functions as narrator of the game's events, keeps track of game time, prompts the comments of the color announcers, reiterates the score, modulates the passage of time, and may lead into commercial breaks.

Player Reference Model: Defines the ideal behavior of a DVD (compliant) Player.

Player: Embodiment of a DVD decoder system which executes the Navigation system and performs all decoding from the channel layer at least up to the track buffer layer. In future, external MPEG decoders may perform the actual video and audio reconstruction, but copyright issues currently prevent this.

Plot: The basic story line of a dramatic production.

Plug and Play: Ability of computer operating system to automatically detect and adapt to a new component.

PMMA: See Polymethylmethacrylate.

POD: The group of commercials and announcements in a break during programming.

Polar Pattern: The range of sensitivity surrounding a microphone as plotted in a two- or three-dimensional drawing.

Polarity Reversal: In video, the reversal of the gray scale tones (white becomes black, etc.) or colors (colors turn into their complements). In audio, a reversal of wires resulting in a positive charge becoming negative and the negative becoming positive.

Polarity: The positive or negative characteristics of an image or electrical signal.

Polarizing Filter: A filter (often adjustable) that reduces or eliminates reflections from glass, water and shiny surfaces. Also used to dramatically intensify contrast between the sky and clouds.

Pole Cat: A pole with spring-loaded, telescoping sections placed between the floor and grid pipes and used to attach scenery and set pieces.

Polish: The writing of changes in dialogue, narration, or action, but not including a rewrite.

Polyester Base: Polyester base is a very durable type of film that is virtually impossible to rip. Some people claim that it is harder to splice, but that is more a matter of getting used to the technique. Significantly, it cannot be cement spliced, making it impractical as original material (also, its durability could spell disaster for the delicate mechanism of a camera in the event of a jam). However, its durability makes it very advantageous for release prints.

Polygon Budget: A predetermined amount of polygons that makes the engine run with acceptable frame rate. Going "over budget" will negatively impact the frame rate. For example, a character may have a polygon budget of 800 polygons for the artist to work with.

Polymethylmethacrylate (PMMA): A clear acrylic compound used in laserdiscs and as an intermediary in the surface transfer process (STP) for dual-layer DVDs. PMMA is also sometimes used for DVD substrates.

Polysemy: Literally, many meanings. Refers to television's ability to communicate contradictory or ambivalent meanings simultaneously.

Pool-hall Lighting: Lighting from a single source, usually hanging in the middle of the set and visible in the scene.

POP (Picture Outside Picture): A feature of some widescreen displays that uses the unused area around a 4:3 picture to show additional pictures.

Pop Filter: A screen placed over a microphone that reduces the effect of speech plosives and wind.

Pop: Microphone distortion caused by speaking certain consonants (especially "p") into a microphone placed too close to the mouth.

Ports: Openings in a microphone case designed to achieve a specific pickup pattern and frequency response. Also, connections on a computer used for plugging in accessories.

Position: The location of commercials within a specific program.

Positioner: A video effects control which can move an inserted key source to any position in background video.

Positive Scratch: The black image on a print of a scratch on the positive from which the print was made.

Possible Future Impact: A measures of newsworthiness relating to the possible consequences to an audience of events outlined in a story.

Post Buy Analysis: An evaluation based on the audience viewing levels at the time the announcement ran. A post buy analysis is used to determine the effectiveness of an advertising buy. Also called Stewardship Reports.

Post Production: At the point where the First Unit has completed live action work on a media program, movie or television program, and the work to be done consists of non-real time activities such as editing, sound mixing, compositing, image manipulation, image creation, or any alteration of the media product that leads to the final product a production process is said to be in Post Production.

Post: Short for Postproduction.

Post-command: In DVD-Video a navigation command to be executed after the presentation of a Program Chain (PGC) has been completed.

Posterization: Special effect in which the picture is reduced to a small number of colors or luminance levels removing any fine gradations of color and brightness resulting in an oil painting effect.

Postproduction Switcher: A video switcher which is capable of interfacing with two or more videotape machines and designed to meet the specific needs of editing.

Post-recording: Any audio work done after the main taping has been completed.

Post-roll: To run a VTR a defined distance beyond an edit point to check the pictures that follow.

Post-synchronized Sound: Sound added to images after they have been photographed and assembled. This is commonly called Dubbing.

Pot: Potentiometer, a kind of variable resistor. Usually found behind volume and tone control knobs and graphic equalizer sliders, also commonly used as the screwdriver adjustments in back of a TV set to fine tune convergence, pincushion distortion, etc.

POV Shot: Point of View Shot. A shot from the perspective of one of the characters, as if the audience were seeing the scene from their eyes. It is often important to get a Reaction Shot to establish that any given shot really is a POV.

Power Zoom: A lens with a servo motor that makes possible electrically controlled zooms.

PPM: See Peak Program Level Meter.

PPV: See Pay Per View

Practical: Any photo flood-type of bulb, used within the shot, in a household lamp or otherwise visible. The term practical is sometimes used interchangeably with photo flood, even though it specifically refers to a light used in the shot. It is also a prop that will actually be used in a scene, as opposed to those that are meant to be seen only. For example, a light that can be switched on or off by an actor.

Preamplifier: An electronic device that boosts extremely weak signal voltages, such as those from microphones or mag heads, to a level that is usable by power amplifiers.

Pre-Blacked: A videotape which has already had a control track, usually with SMPTE encoded time code, but without any picture or sound. This is done to facilitate the video editing or assembly process.

Pre-command: In DVD-Video a navigation command to

be execute before the presentation of a Program Chain (PGC) has been started.

Preemption: When a regularly scheduled program is replaced by special programming, the regular program is said to have been preempted.

Pre-Enhancement: In many situations, video losses can be anticipated, allowing signal pre-compensation in a way that partially corrects for the losses.

Preferred Reading: In cultural studies, the interpretation of the text that is stressed by the text itself. Marxists presume this reading to align with the dominant ideology.

Pregap: More accurately known as Pregap 1 or Pregap 2. Pregap is the area before a CD-ROM track.

Premastering: The process of preparing data in the final format to create a DVD disc image for mastering. Includes creating DVD control and navigation data, multiplexing data streams together, generating error-correction codes, and performing channel modulation. Often includes the process of encoding video, audio, and subpictures.

Premiere: The first official public screening of a movie, marking the opening. The affair is often a gala event attended by the filmmakers, stars, and other celebrities.

Premise: The basic idea for a story often taking the form of a question or a problem.

Premium Channels: Program services purchased by cable subscribers for an extra fee. An all-movie channel such as HBO is an example of a premium channel.

Preproduction: Arrangements made before the start of filming. This can include script editing, set construction, location scouting, and casting. See also Production.

Prequel: A movie that presents the characters and/or events chronologically before the setting of a previously filmed movie. See also series, serial.

Preroll: 5 to 7 seconds of camera running time before a shot can be used. In editing, this refers to a similar amount of automatic backspacing the edit decks perform to insure a stable edit.

Prescoring: Recording of music or other sound prior to the shooting of the picture which is to accompany it. The most common usage is in animated film.

Presence: The subjective feeling of special "closeness" a listener has to a performer, typically as achieved through a specific balance of frequencies within a reproduced audio signal.

Presentation Data: DVD-Video information such as video, menus, and audio which is presented to the viewer.

Preset Monitor: A video display allowing sources to be previewing and adjusted before use. Also called Preview Monitor.

Pressure Plate: Part of the internal workings of a camera, the pressure plate is located on the other side of the film from the gate. It is a smooth, spring-loaded plate that holds the film on the film plane and acts as a brake, helping to hold the film steady while it is exposed.

Pressure Zone Microphone (PZM): A microphone mounted on a reflecting surface. Sound waves build up a pressure zone within the microphone housing before being transferred to the microphone. Often used for group discussions around a table.

Preview (edit): Going through all the actions of the edit, without recording the result, to check it is correct. For VTR-based editing this means do everything except allow the record machine to be switched to record. Previews are use to judge and grade images or scenes at a point in time where they can be modified before they are committed to their final form. When correct, the procedure must be repeated and the result recorded. In many cases using a true random access disk-based system, once a satisfactory result is seen there is no need to perform the edit again.

Preview Bus: A processor function allowing the operator to select any incoming video source for viewing prior to actual use. Typically, each signal can be previewed on its own monitor. This is an effective method to check work before going "on the air."

Pre-visualization: A term used to describe the process of actually looking at a representative image to be used in a media program, television program or movie. A pre-visualization may include stand-in actors, computer generated elements that are planned to be photographed live, or live photographed images that are planned to be computer generated. Pre-visualization is typically performed at at a point in time where experimenting with the images is more cost effective than the actual manner in which an image would be committed to a media.

Primary Colors: The basic colors used in TV and video systems of red, green and blue. See also RGB.

Primary Delivery Control: Information sent during the vertical blanking interval using teletext to control VCRs in Europe. The specification is ETS 300231.

Prime Lens: A prime lens is one with a single focal length, wide, normal or telephoto, as opposed to a Zoom Lens, which has a variable focal length. They often come in a set of different focal lengths. Prime lenses tend to be sharper, faster and will often focus closer than zoom lenses.

Prime Time Access Rule: An FCC rule forbidding Network Affiliates from carrying more than three hours of Network Programs and Off Network in the four hours starting at 7PM Eastern and Pacific time, 6PM Central and Mountain time. The FCC has recently reversed this regulation.

Prime Time: The most expensive airtime since the audience potential is the greatest. Usually 8 PM-11 PM Monday through Saturday and 7 PM-10 PM on Sundays. In the Central and Mountain Time Zones, prime time starts and ends an hour earlier.

Prime: The video signal or its components are correctly

referred to by the various terms Y, R-Y, G, Cb, etc. without the apostrophe or "prime symbol" when the signal has not been modified to compensate for the gamma of the picture tube. The prime symbol is added, as in Y', Cb', etc. to refer to a video signal after gamma compensation.

Principal Photography: The filming of major or significant components of a movie which involve lead actors.

Print Spooling: This process lays down the image of a printed report on the disk file instead of going directly to the printer. Because of the rapid speed of internal processing, the computer can create the image of the report on the disk faster than feeding each line to the printer. The print-spool program then prints the report from the spool file on the disk, while the operator and the processor proceed to the next job(s).

Print Stock: Film used by the lab for making copies (prints). It is usually of a longer pitch than camera stock so as to be smoothly sandwiched against the camera stock on the printing machine. It is also much slower (with an A.S.A. of about 12) than camera stock, as light is less of a problem in printing than it is when it is being focused through a lens in a camera.

Print: Projectable version of a movie, usually consisting of one or more reels.

Printer's Sync: This is the offsetting of sound 26 frames earlier than picture, corresponding to the distance between the sound reader and the gate of the projector. To be in sync on a projector all prints are lined up in printer's sync. Usually the lab lines up the sound and picture in printer's sync, putting the beep on the track 26 frames earlier than the "2" in the Academy Leader. This is known as pulling up the sound. If there was some reason for you to line up the sound yourself, it is very important to label the sync mark "printer's sync" so that the sound is not accidentally pulled up twice.

Print-through: The undesirable transfer of an audio or

video signal from one layer of tape to another on a tape reel.

Prism Block: See Beam Splitter.

Prism Lens: A special-effects device that attaches to the front of the camera lens and breaks up the primary image into multiple images.

Prism Shutter: A device used on many film viewers, editing machines, and some high-speed cameras, consisting of a rotating prism of four or more sides through which the viewer light passes as film is pulled continuously through it.

Privacy: An individual's—especially a private citizen's—legal right to be left alone and not badgered by news people. Privacy includes intrusion on seclusion, the public disclosure of private facts, false light and commercial appropriation.

Private Television: Non-commercial television that includes corporate, educational, medical and religious television production.

Privilege: A legal term relating to a degree of freedom that certain types of communication have from libel or slander actions. Certain political and legislative communication can be broadcast without being actionable under slander or libel.

Problematic Fit: In the study of television stars, a complete mismatch of a particular role's characteristics with a star's polysemy.

Proc Amp: See Processing Amplifier.

Process Shot: A shot in which "live" foreground action is photographed against a background image projected on a translucent screen.

Processing Amplifier: Video image processor that boosts video signal's luminance, chroma, and sync components to correct such problems as low light, weak color, or wrong tint. Also called Proc Amp.

Producer: The chief of a movie production in all matters save the creative efforts of the director; raising funding, hiring key personnel, and arranging for distributors. In television, usually a current or former writer who has successfully written for a number of years as a staff member on a show and is now responsible for the creative aspects of the show.

Product Placement: The appearance of a trademarked product (e.g., Budweiser beer or Apple computers) in a program-when the sponsor pays for such placement.

Product Protection: The limits of protection granted by a television station against products advertised by competitors.

Production Accountant: The person responsible for managing finances during the production.

Production Assistant: A person responsible for various odd jobs, such as stopping traffic, acting as couriers, fetching items from craft service, etc. Production Assistants are often attached to individual actors or filmmakers.

Production Bonus: Cash bonus given to the writer of a screenplay who receives shared or sole "Screenplay by" or "Written by" credit when the screenplay is turned into a film.

Production Breakdown: The division of the script into individual scenes and shots so that an efficient shooting schedule can be planned.

Production Buyer: A person who purchases supplies, equipment, and property necessary for a production.

Production Company: A general term for a company that is associated with the making of a movie.

Production Date: Refers to the phase of movie making during which principal photography occurs.

Production Designer: An artist responsible for designing the overall visual appearance of a movie.

Production Illustrator: A person responsible for drawing the storyboards and anything else that needs to be drawn during the production of the movie.

Production Line: Wired or wireless headset intercommunication link between production personnel.

Production Manager: Reporting to the film's producer, this person is responsible for the practical matters such as ordering equipment, getting near-location accommodations for the cast and crew, etc.

Production Report: A daily report of actual progress versus the production schedule. Includes dope sheets, continuity reports, and call sheets, as well as extensive notes regarding on-set happenings, activities of the cast and crew, and explanations of unexpected events.

Production Schedule: A detailed plan of the timing of activities associated with the making of a movie, of particular interest to production managers. See also production report.

Production Secretary: Secretary to the production manager.

Production Sound Mixer: The head of the sound department on the set. They are responsible for the process of recording all sync dialog and sync sound effects in a scene. The Production Sound Mixer has a number of duties: selection and operation of the microphones, and recording equipment used on the set, directing the boom operator, combining the sound of multiple microphones used to capture dialog and effects on a set, recording sound ambiance and room tone for all scenes, and wild track that will aid the editor and sound mixer in matching the different sound takes in a scene for smooth sound transitions.

Production Sound: This is the sync sound, or any other sort of wild track or room tone that was recorded at the shoot. The term is used in sound editing to distinguish between added backgrounds and effects and those from the shoot.

Production Switcher: As opposed to a editing switcher or a master control switcher, a switcher designed to work in conjunction with productions originating form a control room or remote van. professional video equipment A term which covers industrial and broadcast-quality equipment.

Production: In the movie industry, this term refers to the phase of movie making during which principal photography occurs. Popularly, however, "production" means the entire movie project. See also pre-production and post-production.

Profile: In MPEG-2, profiles specify syntax and processes such as picture types, scalability, and extensions.

Profile: Side view of a person.

Program (PG): In a general sense, a sequence of audio or video. In a technical sense for DVD-Video, a group of cells within a Program Chain (PGC).

Program Analysis: Studying the effectiveness (audience reach, tune-out factors, etc.) of programming, usually to determine the rates for advertising during the programs, and to see if there should be a change in formats or programming.

Program and System Information Protocol: See PSIP.

Program Audio: Final audio mix intended for recording, broadcast or distribution.

Program Automation: An automation system that automatically switches programs and spots according to a previously entered schedule.

Program Bus: The master bus on a video switcher which determines the output signal of the switcher. Similar to the preview bus in concept except that the resulting output is the final signal which goes "on the air."

Program Chain (PGC): In DVD-Video, a collection of programs, or groups of cells, linked together to create a sequential presentation.

Program Log: See Log.

Program Proposal: A written outline or summary of a proposed production. In dramatic productions they are usually referred to as treatments. See also Production Outline and Treatment.

Progressive Scan: A video scanning system that displays all lines of a frame in one pass. See also Non-interlaced.

Projection Leader: A short length of film having standard markings on it, used to enable projectionists to make instant changeovers from one projector to another.

Projection Sync: Same as Printer's Sync.

Projection Television: An optical and electronic system for creating a large-screen video display.

Projection-Contrast Original: An original reversal film which is designed to have normal contrast when projected.

Projector: A device for displaying a reel of a movie on a screen, either for a screening or a back projection.

Prominence: A measures of newsworthiness relating to how important or well known the people, places or things in the story are to an audience.

Promo: A promotional announcement, usually for the television station, but it can also promote client participation in an event (i.e., a television station promo might mention the evening news with it's award winning news team while a client promo might mention that the advertiser will be giving away television station t-shirts at the ball game.

Prompter: An electrical device for displaying a script that can be read by talent during a production. Also referred to as Teleprompter. This term can also refer to a person who supplies actors with the correct lines from the script if they forget.

Prop: Any movable item used on a theater or film set.

Property Assistant: The person responsible for the

placement and maintenance of props on a set.

Property Master: A person responsible for buying/acquiring any props needed for a production. Contrast with set dresser.

Prosthetic Appliances: Makeup that requires gluing additions (prosthetic appliances) made of a material such as latex or gelatin to an actors skin. May be applied by someone other than the designer. See also Squib.

Prosumer Video Equipment: A combination term denoting equipment which lies between professional and consumer-quality video equipment.

Protection Master: A high quality copy of the master tape.

Protection: A backup copy of a recording made in case the primary recording has problems.

Protocol: Set of "rules" defining exchange of data, including timing, format, sequencing, error checking, etc.

Proxemic Patterns: Spatial patterns of the characters in a frame.

Proximity Effect: The exaggeration of low-frequency response associated with most microphones when they are used at very close distances. Some microphones have built-in adjustments that can compensate for proximity effect.

PSA: The abbreviation for Public Service Announcement, which is a spot containing information of interest to the general public, and aired by the television station at no charge. An example of a PSA would be a spot informing the public of free blood-pressure check-ups on a certain day.

Pseudo 3-D System: A computer-based electronic graphics system capable of being used to do drawings which, rather than appearing flat, seem to have three dimensions.

Pseudo Color: A term used to describe a technique that applies color, or shows color, where it does not really exist. We are all familiar with the satellite photos that show temperature differences across a continent or the multicolored cloud motion sequences on the nightly weather report. These are real-world examples of pseudo color. The color does not really exist. A computer uses a lookup table memory to add the color so information, such as temperature or cloud height, is viewable.

Pseudomonologue: Type of interview in which the interviewer and his or her questions are not evident in the text; only the interviewee's answers are included.

PSIP (Program and System Information Protocol): Pronounced "P-SIP," it's a part of the ATSC digital television specification that enables a DTV receiver to identify program information contributed by content providers and use it to create sophisticated electronic program guides.

Psychoacoustic: See Perceptual Encoding.

PTAR: See Prime Time Access Rule.

PTP: See Parallel Track Path.

PTT Menu: In DVD-Video, a menu used to access specific Part of Title (PTT) in a Video Title Set (VTS). Usually referred to as a Chapter Menu.

Public Access Programming: Cable channels or CATV programs designed to serve the interests of the community.

Public Broadcasting Service (PBS): A non commercial, contribution funded television network which offers primarily news, cultural, and children's education programming (i.e., Mac Neil Lehrer Hour and Sesame Street).

Public Domain: The state in which the creator of a work loses the copyright on it through the passage of the copyright period, failure to renew the work, or problems with the original registration of the work with the copyright office.

Public Property: Property, generally owned by a government, which does not have the legal access limitations inherent in private property. Property which can be freely used by news reporters and photographers.

Public Service Announcement: See PSA.

Publicity Assistant: Assistant to the publicity director.

Publicity Department: The section of a production's crew responsible for promoting a movie. Individual positions within in this department include: publicity director, publicity assistant, and stills photographer.

Publicity Director: Chief of the publicity department, responsible for promoting the movie through the media.

Publicity: A type of media text (e.g., an unauthorized biography) that presents information outside the control of the star and his or her representatives.

PUH (Pickup Head): The assembly of optics and electronics that reads data from a disc.

Pull Down: A transfer of sound slowed down from film speed, 24 film frames per second, to video speed, 29.97 video frames per second, which is the equivalent of 23.98 film frames per second. This must be done to line it up with a video transfer of picture when transferring sync sound to video.

Pull Focus: Shifting focus from one part of a scene to another to redirect viewer attention.

Pull Processing: Pull processing is a special type of processing where the film is developed for a shorter time than normal, usually to make up for intended overexposure.

Pull Up: The process of offsetting the sound 26 frames ahead of picture when making a print. Pull Ups, as a noun, are transfers of the first 26 frames of sound from a reel that are spliced onto the outgoing sound of the previous reel so that sound is not lost when the film is printed with the sound pulled up, since 26 frames of sound are cut off when reels are joined. It is also a term used to describe a transfer of the sound from a video, sped up from video speed, 29.97 video frames per second, which is the equivalent of 23.98 film frames per second, to film speed, 24 film frames per second. This must be done when the optical track is made after having mixed in video.

Pull-down Claw: Part of the camera movement which advances the film from the exposed frame to the next unexposed frame while the camera's shutter is closed.

Pulse Code Modulation (PCM): Method of recording high quality, digital sound on 8mm and Hi8 tapes.

Pulse Width Modulation (PWM): PWM is a way of digitally encoding analog signal levels, which allows for digital control of analog circuits. This, in turn, helps to reduce cost, size, heat, and power consumption in devices such as consumer audio hardware.

Punch: A device for punching a hole in film leader to locate a starting point for editorial or printing synchronization. Also refers to Specular Light.

Puppet Film: An animated film in which inanimate objects or figures are manipulated and photographed frame by frame in order to make them appear to move when the film is projected.

Pure Film: A type of experimental film that explores the purely visual possibilities of cinema rather than narrative possibilities. Also called Pure Cinema.

Pure Metal Tape: A recording tape capable of a high signal-to-noise ratio, high-frequency response and excellent record-playback efficiency.

Push Processing: Push processing is a special type of processing where the film is developed for a longer time than normal, usually to make up for intended underexposure. It should be noted that only entire rolls can be pushed, not individual scenes. Pushing film will add some contrast and graininess.

Push-off Wipe: An electronic video effect in which one image replaces another by appearing to push it off the screen.

Push-Pull: See Radial Tracking.

PWM: See Pulse Width Modulation.

Pylon Triangular: A set piece resembling a pillar.

Pyrotechnician: A member of the crew with expertise in fire or explosions.

Pyrotechnics: Special effects involving explosions.

PZM: See Pressure Zone Microphone.

Q

Q Signal: One of the two color signals, containing yellow and violet components to which the human eye is relatively insensitive.

Q: See IQ.

QAM: See Quadrature Amplitude Modulation.

QCIF (Quarter Common Interface Format): This video format was developed to allow the implementation of cheaper video phones. The QCIF format has a resolution of 176 x 144 active pixels and a refresh rate of 29.97 frames per second.

Quad Chroma: Refers to a technique where the sample clock is four times the frequency of the colorburst. For NTSC this means that the sample clock is about 14.32 MHz (4 x 3.579545 MHz), while for PAL the sample clock is about 17.73 MHz (4 x 4.43361875 MHz). The reason these are popular sample clock frequencies is that, depending on the method chosen, they make the chrominance (color) decoding and encoding easier.

Quad Microphone: A four-element microphone designed to detect quadrants of sound around a full 360-degrees. See Quadraphonic Sound.

Quadraphonic Sound: A system of sound reproduction that goes beyond stereo and attempts to record and playback a 360-degree sound perspective. See Quad Microphone.

Quadrature Amplitude Modulation (QAM): The modulation of two carrier components, which are 90 degrees apart in phase.

Quadruplex: The original system of tape recording in which four rotating video heads record and play back the image. The tracks on the tape are nearly vertical as opposed to those in longitudinal recording. Also called a Quad.

Quality Control Form: See Discrepancy Report.

Quality: A dimension of light relating to its hardness or softness. See Coherence.

Quantization Error: The difference between the actual value of the analog signal when it is sampled and the resulting digital word value.

Quantization Levels: The predetermined levels at which an analog signal can be sampled as determined by the resolution of the analog-to-digital converter (in bits per sample); or the number of bits stored for the sampled signal.

Quantization Noise: The deviation of a signal from its original or correct value which results from the quantization process. This is the inherent uncertainty introduced during quantization since only discrete, rather than continuous, levels are generated. Also called Quantization Distortion.

Quantization: The process of converting a continuous analog signal into a set of discrete levels (digitizing).

Quarter Apple: See Apple Box.

Quarter Common Interface Format: See QCIF.

Quarter Inch: A reference to the standard width magnetic audio tape which is used to record film production sound.

Quarter-track: An audio tape recording system that divides the tape into two sets of stereo tracks. Two tracks are recorded as the tape moves in one direction and two tracks are recorded as the tape moves in the opposite direction.

Quartz Lamp: Widely used in television and film production, an incandescent lamp with a number of internal improvements over standard lamps, including a consistent color temperature and uniform light output over its life. Also referred to as a Tungsten-halogen Lamp.

Quartz Light: Can also be called halogen light or tungsten light. A quartz light is a very bright type of light that uses a tungsten filament that is contained in a quartz envelope. The color temperature will be a fairly consistent 3,200K. They can get very hot when in use.

Quick Release Shoe: The part of the quick release that attaches to the camera is called the quick release shoe, and is inevitably worth double-checking, as they frequently stray away the tripod when left behind on the camera.

QuickTime: A digital video software standard developed by Apple Computer for Macintosh (Mac OS) and Windows operating systems. QuickTime is used to support audio and video from a DVD.

Quota Quickies: Low budget, short subjects made in the UK in the 1930s. The Cinematographic Films Bill, passed in April 1927 stipulated that all UK cinemas would have to include a proportion of British films in their programs. The percentage varied, rising to 20% at one time. The "Quota Quickies" were generally of a quite low quality but there are some gems hidden in their midst, notably those by Michael Powell.

QXGA: A video graphics resolution of 2048x1536.

R

R Speeds: R speeds give you a "polygon per scene" measurement for the map. To enable them in the map, type "r_speeds 1" at the console (without the quotes). To disable, type "r_speeds 0." This command is usable in both hardware accelerated and software modes.

R/F Connectors: Output for coaxial cable (the stuff cable companies use). R/F connectors will not provide stereo from the satellite receiver.

R: A certificate issued by the MPAA indicating that persons under the age of 16 would only be admitted when accompanied by an adult. The age was later raised to under 17 years old, and varies in some jurisdictions. See also NC-17, PG-13.

R_drawflat: A design tool that shows how the level's brushes have been split up by the compile tools. To enable this in the map, type "r_drawflat 1" at the console (without the quotes). To disable, type "r_drawflat 0." This command is usable only in software mode.

R_draworder: A design tool that enables the engine to draw the level in reverse order (from closest to farthest) to show you how much the engine is seeing of the level. To enable this in the map, type "r_draworder 1" at the console (without the quotes). To disable, type "r_draworder 0." This command is usable only in software mode.

RA: See Reverse Angle.

Rack Focus: A shot where focus is changed while shooting. Unlike a Follow Focus shot, a rack focus shot is usually done not from the necessity of keeping someone in focus but to shift attention from one thing to another.

Rack: A frame carrying film in a processing machine. Often used to refer to frame edit alignment in which the projected film remains properly framed on the screen (in rack).

Radial Tracking (RT): Radial Tracking (RT) is an indication of how accurately the laser pick-up of a CD player can locate and follow the track on a particular disc. The RT value is inversely proportional to the depth of the pits.

Radio Frequency (RF): Electromagnetic radiation within a certain frequency range. Radio and television transmitters both utilize RF energy to broadcast their signals.

Radio Microphone: Transmitter microphone or wireless microphone. A microphone connected to a small radio transmitter, used in situations where cables would be cumbersome or impossible to use. FCC regulations are stringent in the use of transmitter microphones.

RAID (Redundant Array of Inexpensive Drives): A system of teaming together numerous computer hard drives, generally for storage in nonlinear editing systems.

Rails: Tracks for camera mounts laid on the ground (resembling small railroad tracks) which allow for smooth follow shots for cameras.

Rain Fade: The loss of signal from the satellite during a heavy rain. This happens more or less to all DBS systems.

Rainbow Filter: Special effect optical filter that creates multicolor streaks of light around spectral highlights and lights.

RAM (Random-Access Memory): Generally refers to solid-state chips. In the case of DVD-RAM, the term was borrowed to indicate ability to read and write at any point on the disc.

RAMbo Drive: A DVD-RAM drive capable of reading and writing CD-R and CD-RW media.

Ramping: Related to the tendency of zoom lenses to have reduced light transmission capabilities as the lens is zoomed in and the mechanical compensation measures built into the lens which automatically open up the iris as the lens is zoomed in and close it down as the lens is zoomed out.

Random Access: The ability to jump to a point on a storage medium.

Random Interlace: Scanning associated with industrial (non-broadcast) sync. Video scanning method in which the horizontal and vertical scan controls run independently of each other and have a somewhat random relationship.

Range Extender: An optical attachment to a lens (or an internal device) that increases focal length.

Rank: A respectable and commonly used brand of Telecine machines. The word is sometimes used interchangeably with telecine in much the same way as "Steenbeck" is used in place of "flatbed." Also a shortened term which refers to a "Rank-Cintel Flying Spot Scanner." This is a telecine device which transfers a film image into electronic signals which are then recorded on to videotape.

Raster: The pattern of parallel horizontal scan lines that makes up a video picture.

Rate Card: A television station's printed price list, rate cards are issued at intervals (e.g., quarterly) and show the prices charged for buying time for various programs, times of day, or packages.

Rate Conversion: The process of converting from one sample rate to another. The digital sample rate for the component format is 13.5 MHz. For the composite format it is either 14.3 MHz for NTSC or 17.7 MHz for PAL. Often used incorrectly to indicate both resampling of digital rates and encoding/decoding.

Rate Protection: An agreement guaranteeing an advertiser a continuous set rate unaffected by any rate increase.

Rating: Usually stated in conjunction with Share, a rating is the percent of all people in a demographic group in a survey area that are tuned in to a specific television station. The number is derived by dividing the number of people tuned in by the total population of the survey area. Example: There are 1.9 million people in the Denver Metropolitan area. 12% of the people that own TV's are tuned in to Channel 4. Channel 4 has a 12 rating. See also Share.

Raw Photage: Pre-edited recordings, usually direct from camera.

Raw Stock: Unexposed film.

Raw VBI Data: A technique where VBI data (such as teletext and captioning data) is sampled by a fast sample clock (i.e. 27 MHz), and output. This technique allows software decoding of the VBI data to be done.

RC Time Code (Rewriteable Consumer): A time code system, available on 8mm and Hi-8 formats only. The code can be added either before or after video recording without affecting the video or audio.

RCA Connector: A type of connector used on all consumer VCRs and camcorders to carry the standard composite video and audio signals.

RCA Plug and Jack: A one conductor plus ground plug and jack connector introduced by Radio Corporation of America, now part of Thomson Electronics. The jack is a metal stud a little larger around than a pencil (about 3/8 inch) with a center hole that a small lollipop stick would fit (about 1/8 inch). The matching plug has one fat (1/8 inch) center pin that represents the live conductor, and a shell that presses onto the jack stud and that is almost always grounded. Normally the center pin is connected to the center conductor of a coaxial cable and the shell is connected to the shell or shield of the cable.

RCD: See Remote Control Device.

Reach: The percentage of different households or demographics that were exposed to a media schedule at least one time. Since reach is an unduplicated measurement of all exposures to a household/demographic, it can never exceed 100%. Expressed as an absolute number, reach is referred to as net impressions or net audience.

Reaction Shot: A shot of someone looking off screen. Used either to lead into a POV Shot or to show a reaction right after a P.O.V. shot. A reaction shot can also be a shot of someone in a conversation where they are not given a line of dialogue but are just listening to the other person speak.

Read Throughs: The somewhat informal sessions attended by the director and actors in which scripts are initially reviewed and rehearsed to decide on dramatic and production emphasis.

Reader: Individual who reads scripts and writes down synopsis of the plotline, offering positive or negative comments (the process is called "providing coverage"), which assist studio execs or interested parties if script is worthwhile.

Reading Period: Period after a writer has been hired to write an assignment that a hiring body will review. This body will give suggestions and decide whether to pick up the option to have the writer produce further work.

Reading: The viewer's active interpretation of a text: whether written (e.g., a book) or visual (e.g., a television program or film).

Read-modify-write: An operation used in writing to DVD-RAM discs. Because data can be written by the host computer in blocks as small as 2 KB, but the DVD format uses ECC blocks of 32 KB, an entire ECC block is read from the data buffer or disc, modified to include the new data and new ECC data, then written back to the data buffer and disc.

Real Time Control Protocol: See RTCP.

Real Time Counter: A display showing hours-minutes-seconds of tape that has been recorded (elapsed time), or how much tape remains.

Real Time Protocol: See RTP.

Real Time Transport Protocol: See RTSP.

Real Time: The actual time during which video recording

occurs, distinguished from the tampering of time via editing.

Realism: A style of filmmaking which endeavors to depict physical reality much as it appears in the everyday world. Typical realistic techniques include the prominent use of long shots, eye-level camera angles, lengthy takes, naturalistic lighting and sound effects, and unobtrusive editing. See also Expressionism.

Realistic Set: A production setting which appears (from the camera's viewpoint) to be authentic.

Rear Projection: Screen used for video effects onto which images are projected from the rear. When actors are photographed form the front, it appears as if they are part of the projected scene.

Recans: Leftover film that was loaded into a magazine but (unlike a Shortend) not even partially shot, and then loaded back in the film can. Basically, it is a roll a film that has been opened, but not used.

Receiver: Any device capable of demodulating an RF signal, such as a radio, tuner, or television set.

Reclocking: The process of clocking the data with a regenerated clock.

Reconciliation: The process of balancing what spots aired and what spots were ordered is called reconciliation. Program events and commercial announcements are sometimes not broadcast according to the schedule on the system-produced log. The reason may be that a regularly scheduled program was preempted or perhaps a human or mechanical error forced a change in the schedule. To reflect the actual sequence of events, manual revisions are made to the log as needed throughout the broadcast day. The differences between the scheduled and aired logs must then be reconciled so that advertisers will be billed properly.

Record Head: The electromagnetic device that places an audio or video signal onto a recording medium (such as videotape) for later playback.

Record VTR: The video recorder that produces the final edited master by recording the edits from one or more source machines.

Recorder: Any device that converts an electronic signal to a magnetic pattern in the oxide coating of a magnetic tape.

Recoupment: Gross funds from a film that are required to pay off negative costs, overhead, ongoing distribution fees, interest, financing and distribution costs, and, in appropriate cases, payment of gross participations.

Rectangular Pixels: Pixels that are not "square pixels" are "rectangular pixels."

Red Book: The document first published in 1982 that specifies the original compact disc digital audio format developed by Philips and Sony.

Redhead: An open faced 1K lighting unit. Also known as a Microphonekey.

Reduction Print: An optical reduction of a film from one gauge to another, such as 35mm to 16mm.

Reed-Solomon: An error-correction encoding system that cycles data multiple times through a mathematical transformation in order to increase the effectiveness of the error correction, especially for burst errors (errors concentrated closely together, as from a scratch or physical defect). DVD uses rows and columns of Reed-Solomon encoding in a two-dimensional lattice, called Reed-Solomon product code (RS-PC).

Reel (Desk): An area of workspace on the Edit Desk of a Quantel editing system into which video and audio clips are placed.

Reel ID: An alphanumeric label assigned to source videotapes to keep track of reference video and audio throughout the editing and conforming process. Reel ID forms part of Edit Decision Lists (EDL).

Reel: A metal or plastic spool for holding film, either for projection or editing. In 35mm a reel is 1,000 feet of film (or usually a little less). Also known as a Single Reel.

Reel-to-reel: In contrast to a cartridge or cassette tape format, a tape format that depends on two separate reels: a supply reel and a take-up reel.

Re-establishing Shot: A long shot which once again positions the character(s) within the environment of the scene, helping to re-establish character and/or setting; also used as a transitional device.

Reference Black: The darkest portion of the video picture, generally with a reflectance value of 3 percent.

Reference Picture: An encoded frame that is used as a reference point from which to build dependent frames. In MPEG-2, I pictures and P pictures are used as references. Also known as a Reference Frame.

Reference Player: A DVD player that defines the ideal behavior as specified by the DVD-Video standard.

Reference Video Signal: A video signal which contains a sync signal or sync and burst signals, used as a reference for synchronization of video equipment.

Reference White: The whitest portion of the video picture, generally with a reflectance value of 60 percent or above.

Reflected Light Meter: A photoelectric device that measures the amount of light reflected from a subject. In determining exposure, reflected light meters assume an average 18 percent reflectance.

Reflected Light: Light reaching the camera after being reflected by a subject. See also Incident Light.

Reflective Light Reading: A reflective light reading measures the amount of light bouncing off the subject. You take a reflective reading with a light meter equipped with a honey-comb or lensed grid. The meter is pointed at the subject, so as to read only the light bouncing off the subject. The other type of light reading is an Incident Light Reading.

Reflector Board: A silver or bright white surface used to reflect light onto a subject. Generally used outside to soften and fill in the light from the sun.

Reflector Spot: A sealed spotlight with a self-contained reflector.

Reflector: Lighting accessory helpful for spreading light as well as filling in shadows. Often made of lightweight reflective metal or poster board covered with metallic material.

Reflex: A viewfinding system in a camera where the image you see in the viewfinder is viewed through the same lens that is used to photograph the image on film.

Reflexive Mode: Type of non-narrative television text which draws the viewer's attention to the processes, techniques and conventions of television production.

Regional Code: A code identifying one of the world regions for restricting DVD-Video playback.

Regional Management: A mandatory feature of DVD-Video to restrict the playback of a disc to a specific geographical region. Each player and DVD-ROM drive includes a single regional code, and each disc side can specify in which regions it is allowed to be played. Regional coding is optional-a disc without regional codes will play in all players in all regions.

Registration Pin: A registration pin is found in the movement certain cameras, such as the Arriflex and the Eclair, and acts to steady the image during exposure.

Registration: The degree to which one frame lines up with the next is registration. A camera with poor registration will create an image that will gently bobble when projected. Projectors too can have good or poor registration (sometimes making it difficult to tell if it was the camera). Good registration is most important for certain types of special effects shots where images are layered and will call attention to themselves if they are gently bobbling out of sync with each other.

Rehearsal Hall: Any area in which a dry rehearsal takes place.

Relational Editing: Editing of shots for the purposes of comparison or for the contrast of content.

Relay Lens: An internal optical device used in some cameras that transports (relays) the image from the lens over an extended path before it reaches the camera's target.

Release Form: Statement signed by an individual that generally frees the creator of the document from any kind of liability.

Release Level: The term used to identify software packages and subsequent updates to those packages is called the release level.

Release Negative: A duplicate negative from which release prints are made.

Release Print: This is a print made after the answer print has been approved. It is not retimed, but struck using the same timing as the final answer print. Because it is not retimed it is generally much cheaper than an answer print. On a big production, these are the prints released to movie theaters, hence the name.

Rembrandt Lighting: See Back Lighting.

Remote Control Device (RCD): A device that allows one to operate a television without directly touching it.

Remote Survey: An evaluation of a field location by key production and engineering personnel so production needs can be ascertained.

Remote Truck: A vehicle containing basic field production gear.

Remote: Any multiple-camera telecast from outside the studio. Television remotes are often live coverage of sports events, usually with a van or mobile home equipped for remote broadcasting.

Render: Computer process of performing the calculations to create detail in a three-dimensional computer-generated scene or special effect.

Rendering Time: The time it takes a DTV computer to compute a wipe or DVE from two video sources, typically 30 seconds to "render" a 30-frame effect. Analog SEGs do this in real time.

Rep Firm: A national company that represents one local television station per market to national advertisers is known as a rep firm.

Repertory: Naturalist performance style in which the actor constructs a performance by selecting particular gestures and spoken dialects.

Replication: The reproduction of media such as optical discs by stamping (contrast with duplication). Replication is also a process used to increase the size of an image by repeating pixels (to increase the horizontal size) and/or lines (to increase the vertical size) or to increase the display rate of a video stream by repeating frames. For example, a 360x240 pixel image can be displayed at 720x480 size by duplicating each pixel on each line and then duplicating each line. In this case the resulting image contains blocks of four identical pixels. Obviously, image replication can cause blockiness. A 24-fps video signal can be displayed at 72 fps by repeating each frame three times. Frame replication can cause jerkiness of motion. Also see Interpolate.

Representational-supportive Set: A production setting, such as a news or weather set, designed solely to support the function of the production.

Rerecording: The process of mixing all edited music, effects and dialog tracks of a film or video production to mono, stereo, multichannel, or whatever audio format is desired for the final print master.

Resampling: The process of converting between different spatial resolutions or different temporal resolutions. This may be based on simple sampling of the source information at higher or lower resolution or may include interpolation to correct for differences in pixel aspect ratios or to adjust for differences in display rates.

Reserved Bytes: 6 bytes in the header of each DVD sector reserved for future use.

Residual Subcarrier: This is the amount of color subcarrier information present in white, gray, or black areas of a composite color video signal (ideally, there is none present). The number usually appears as -n dB. The larger "n" is, the better.

Resolution Chart: A test pattern used to set up and check a camera which shows camera sharpness and the condition of the camera system.

Resolution Independent: A term to describe equipment that can work in more than resolution. For example, most equipment can do film resolution or video resolution, but not both. Resolution independent equipment can work in both.

Resolution: A measurement of relative detail of a digital display, typically given in pixels of width and height. Resolution is also the ability of an imaging system to make clearly distinguishable or resolvable the details of an image. This includes spatial resolution (the clarity of a single image), temporal resolution (the clarity of a moving image or moving object), and perceived resolution (the apparent resolution of a display from the observer's point of view). Analog video is often measured as a number of lines of horizontal resolution over the number of scan lines. Digital video is typically measured as a number of horizontal pixels by vertical pixels. Film is typically measured as a number of line pairs per millimeter.

Resolver: A device that governs the speed of a tape recorder during the transfer to mag, insuring the sound will be in sync with picture. The resolver uses the pilottone as a reference for adjusting the playback speed, hence something can only be resolved if it has been recorded with a properly equipped tape recorder. The Nagra IV has a built-in resolver.

Resolving Power: The ability of a lens to optically reproduce fine detail.

Retransmission Consent: Part of the 1992 Cable Act that requires cable systems to pay a subscriber fee to local broadcast television stations (affiliates and independents) which appear on the cable system. The rule has been responsible for the startup of such new cable networks as ESPN2, FX and America's Talking.

Reveal: To dolly or zoom back a camera to show additional material.

Reverberation Unit: A mechanical or an electronic device that creates controlled sound reverberation or echo. Used for special effects and in postproduction sweetening to balance location presence between different sources of audio.

Reversal Film: A film that is manufactured and processed in such a way as to produce a positive image after exposure.

Reversal Intermediate: A second generation duplicate which is reversed to make it the same type, negative or positive, as the original. It is used for printing in order to protect the original.

Reversal Original: A reversal film designed to be exposed in a camera.

Reverse Action: An optical effect in which the action appears backwards from its chronological sequence.

Reverse Angle (RA): A shot where the camera is placed opposite its position in the previous shot, "reversing" its view of the scene.

Reverse Motion: Action that moves backward on the screen, achieved by reversing film footage during editing or by reverse printing in an optical printer. Also called Reverse Action.

Reverse Shot: A shot from the other side of the previous shot (though preferably on the same side of the 180 Line), such as cutting between two characters talking, a person exiting and entering though a doorway, a reaction shot and POV shot, etc.

Reverse Telecine: The process that converts media initially developed or already transferred to a digitized format back to a film stock.

Reverse Video: A Negative image. Changing from dark characters on a light background to light characters on a light background, for example. Also called Polarity Reversal.

Rewinds: A simple device for winding film, consisting of a crank and a spindle for mounting one or more reels, typically found mounted on either side of an editing bench.

Rewrite: More than a polish, this is considered the writing of significant changes in plot, story line, or interrelationship of characters in a screenplay.

RF Converter: Device that converts audio and video signals into a combined RF signal suitable for reception by a standard TV.

RF Distribution: The process of supplying an RF signal to several devices simultaneously.

RF Interference: An unwanted radio frequency signal that interferes with an audio or video signal.

RF Microphone: A combination microphone and miniature broadcast transmitter which eliminates the need for microphone cables.

RF Modulation: The process of combining a video signal with an RF source so the result can be transmitted to a television or VCR.

RF: See Radio Frequency.

RG59: The coaxial cable that is commonly used for cable TV. If a home already has coaxial cable, it probably is RG59. It is a smaller gauge than RG6.

RG6: The coaxial cable recommended by DBS manufacturers. It is a larger gauge than RG6.

RGB (Red/Green/Blue): The basic components of a color video signal. Using a color encoder, in conjunction with sync information, a complete composite video signal comprising luminance, chrominance and sync can be generated from RGB.

RGB Encoder: Device that converts composite video into RGB video.

RGB Monitor: A type of color monitor with separate inputs for red, green and blue. Normally associated with high-resolution display systems. See also RGB Video.

Rhythm: The timing of speech, music, sound effects, or editing.

Rhythmic Match: Any juxtaposition of images with actions moving at similar rates of speed.

Ribbon Microphone: A type of microphone that makes use of a metal ribbon suspended in a magnetic field. When sound waves strike the ribbon, the resulting vibratory movement of the ribbon creates a minute voltage.

Rich Map Format: See RMF.

Ride Film: A program that is intended on being finally viewed by an audience that is in a simulator. Typically a ride film is a POV (point of view) program such as would simulate riding in a vehicle or experiencing someone's POV.

Ride Focus: See Follow Focus.

Ride Gain: To manually keep the audio or video levels within acceptable limits.

Rider W: Provision of the Artists/Managers Basic Agreement of the Writers Guild of America in which the contract between WGA members and signatory agencies has been pre-negotiated.

Rights Management: The process of monitoring the ownership, distribution and usage of media.

Rim Lighting: Placing a lamp behind a subject, often with little or no front light, to give a glowing edge or outline of light around the subject.

Ringing: In video, closely spaced repeated ghosts of a vertical or diagonal edge where dark changes to light or vice versa, going from left to right. The electron beam upon changing from dark to light or vice versa instead of changing quickly to the desired intensity and staying there, overshoots and undershoots a few times. This bouncing could occur anywhere in the electronics or cabling and is often caused by or accentuated by a too high setting of the sharpness control. See also Ghosting.

Rip: The term used to describe the extraction of an audio track from a CD.

Ripple: Automatic updating of an Edit Decision List after making a change to the list.

Riser: A cylindrical metal device placed between the dolly head and the camera base to raise the camera. Also a pre-built platform used to raise the set, camera, or lights.

Rising Action: Notion of dramatic rhythm in which events in a story build upon one another with increasing momentum.

Rivas: A type of tape splicer which uses perforated splicing tape. Two models exist—one for straight cuts used for picture, and one for slanted cuts used for sound.

RLE: See Run Length Coding.

RMA (Region Management information): In DVD-Video, an 8-bit (1 byte) value stored in the CPR_MAI that indicates in which regions of the world a disc is permitted to play. See Regional Management.

RMF (Rich Map Format): This is Worldcraft's proprietary map format. VisGroups and other information not strictly part of a .map are stored when using this format. It is not compatible with other editors or with the compile tools, and so must be converted to .map format before distributing or compiling.

RMS (Root-mean-square): Effective sound pressure.

Roadblocking: The scheduling of an advertiser's commercial to air at the same time on all television networks to maximize reach.

Robotic Camera: Any one of several types of remotely-controlled, automated camera positioning systems. Most allow for camera pans, tilts, dollies and trucks.

Roll: A lack of vertical synchronization which causes a complete TV picture to appear to rotate upward or downward. Roll also means text or graphics—usually credits—that move up or down the screen, typically from bottom to top. Produced with character generator.

Roll-off: The gradual reduction of frequencies above or below a certain point. Filters which roll off the bass frequencies are often included in unidirectional microphones to compensate for proximity effect.

Roll-over: See Cut-in.

Roll-through: A term that tells the switcher that the length of the spots is equal to the length of black on the tape. Therefore, the program tape can be left running while the spots air. At the end of the break, the program tape will be positioned correctly to run the next segment. Ordinarily, there are fewer seconds of black than the sum of the spots. In this case, the switcher must stop the tape and re-cue it at the beginning of the next segment.

Rolor: Real color depth expressed in terms of human vision and not "bits per pixel."

Room Tone: A recording of the "silence" of a room or any location, to be used to fill in gaps when editing the sound. The silence of a location is really not very silent at all, and the room tone of one location is not a substitute for another, so a sync sound shoot will usually end with the sound recordist asking everyone to be quiet for the recording of 30 seconds of room tone.

Root Menu: Menu used to access other interactive menus

in the Video Tile Set Manager domain, or to make a selection which is not defined by other system menus such as Angle Menu, Audio, Menu, PTT Menu, and Sub-picture Menu.

ROS (Run-of-schedule): An ROS buy usually means a spot can air anywhere during the day or night. Television stations usually offer advertisers ROS schedules at a lesser rate than specified time schedules. This means that the system will fill avails with higher priority spots first, then fill out the breaks with as many ROS spots as it can. Another phrase describing this type of buy is "Best-Time-Available."

Rotary Fader: As opposed to a linear fader, a volume control regulated by rotation.

Rotation: See Horizontal Rotation and Vertical Rotation.

Rotoscope: A device used in animation wherein a single frame from a live-action film is rear-projected onto a light table with a semi-opaque glass in the center; the animator traces the images cast by the film onto a cel; the tracings are rephotographed, resulting in an animated film that duplicates live-action images.

Rough Cut: The edited film, between the stages of being an assembly and a fine cut.

Roundy-round: See Reverse Shot.

Routing Switcher: A device used to select and direct various sources to specific pieces of audio or video equipment.

Royalty: A fee paid for the use of copyrighted material.

RPM (Revolutions Per Minute): A measure of rotational speed.

RS: See Reed-Solomon.

RS-232/RS-422: Computer communication standards used in video for the control of certain video equipment. Computer controlled VCRs, edit controllers, switchers and other studio equipment can commonly be found in professional video studios. Successfully linking two devices, at the very least, requires that they use the same communication protocol.

RS-CIRC: See Circ.

RSDL (Reverse-spiral Dual-layer): See Opposite Track Path (OTP).

RT: See Radial Tracking.

Rule of Thirds: A composition consideration suggesting that a picture appeals most with its primary point of interest appearing off-center. With screen divided into thirds vertically and horizontally, important elements should be targeted wherever imaginary lines cross.

Rumble Filter: Audio filter that eliminates low frequencies associated with such things as wind, hum, and other types of low frequency noise.

Run of the Picture: A cast member whose work may be required any of the days scheduled for principal photography without incurring liability for additional compensation.

Runaway Production: A production done outside of California or outside the United States, generally with the intent of saving money.

Run-length Coding (RLE): Lossless compression method that exploits contiguous samples with same value.

Running Time: The duration of a finished film.

Runout Signal: A few seconds of black at the end of a videotape recording intended to keep the screen in black and the sync stable while other programming is switched up.

Rushes: Prints made immediately after a day's shooting so they can be viewed the following day. Also know as Dailies.

Russian Montage: A style of editing, typical of prominent Soviet filmmakers in the 1920s, which employs dynamic cutting techniques to evoke strong emotional, and even physical, reactions to film images.

R-V Signal: R (red) signal minus Y (luminance) signal; one of the color difference signals.

S

S/N (Signal-to-noise Ratio): The relation of the strength of the desired signal to the accompanying electronic interference, the noise. If S/N is high, sounds are reproduced with less noise and pictures are reproduced clearly without snow.

S/P DIF (Sony/Philips Digital Interface): A consumer version of the AES/EBU digital audio transmission standard. Most DVD players include S/P DIF coaxial digital audio connectors providing PCM and encoded digital audio output.

Safe Action Area: The inner 90 percent of the video frame. Since the outer 10 percent of a broadcast picture is typically cut off by over-scanning, this area is considered safe for most subject matter.

Safety Chain: A metal chain (sometimes a cable) attached to both a light and a lighting grid, designed to keep a light from falling if its clamp comes loose.

Safety: An additional take, done after a successful one, as a backup.

SAG: See Screen Actors Guild.

Sales Projection Report: This report is used as the primary reference of television station time sales for management and sales department analysis. The sales projection report calculates the revenue that will be generated (for the inventory represented) by those contracts with spots to be aired in the date span of the report. Also referred to as Proj.

Sample Rate: The number of times a digital sample is taken, measured in samples per second, or Hertz. The more often samples are taken, the better a digital signal can represent the original analog signal. Sampling theory states that the sampling frequency must be more than twice the signal frequency in order to reproduce the signal without aliasing. DVD PCM audio allows sampling rates of 48 and 96 kHz.

Sample Reel: Contains samples of a person's or company's best video work for the purposes of marketing; a.k.a. demo reel.

Sample Script: Script a writer has created on his own initiative and used to attain meetings for the writer in order to expose him to the entertainment industry.

Sample Size: The number of bits used to store a sample. Also called resolution. In general, the more bits allocated per sample, the better the reproduction of the original analog information. Audio sample size determines the dynamic range. DVD PCM audio uses sample sizes of 16, 20, or 24 bits.

Sample: A single digital measurement of analog information. A snapshot in time of a continuous analog waveform. See also Sampling.

Sampler: Device that converts analog sound into digital information which thereafter can be fully manipulated by a computer or MIDI device.

Sampling Frequency: The number of sample measurements taken from an analog signal in a second, generally expressed in MegaHertz. These samples are then converted into digital numeric values to create the digital signal.

Sandbag: A cloth bag with two chambers filled with sand, used as a weight on the legs of a light stand for additional stability.

SAP: See Second Audio Program.

Satellite News Vehicle (SNV): A field production used to uplink ENG signals to a satellite. Often contain VCRs and basic editing facilities.

Satellite Television Station: A television station that duplicates some or all of the programming of a pilot television station is known as a satellite. It is assigned separate call letters and channel numbers.

Satellite: An orbiting space television station designed to receive and re-transmit audio and video signals from a wide area (footprint) of the earth.

Saticon: A television pickup tube used mostly in industrial television and electronic news gathering.

Saturated Color: A pure color, one that does not have significant amounts of white or black.

Saturation: In terms of television's image quality, the level of a color's purity (or how much or little grayness is mixed with the color).

SBE: Society of Broadcast Engineers.

Scale: The minimum fees prescribed by unions for television talent, usually writers.

Scaling: Altering the spatial resolution of a single image to increase or reduce the size; or altering the temporal resolution of an image sequence to increase or decrease the rate of display. Techniques include decimation, interpolation, motion compensation, replication, resampling, and subsampling. Most scaling methods introduce artifacts.

Scan Converter: Device that changes scan rate of a video signal, possibly converting it from non-interlaced to interlaced mode. Allows computer graphics to be displayed on a standard video screen, for example.

Scan Line: A scan line is an individual line across the display. It takes 525 if these scan lines to make up a NTSC TV picture and 625 scan lines to make up a PAL TV picture.

Scan Rate: Number of times a screen is "redrawn" per second. Computer displays operate at different scan rates than standard video.

Scanner: A device that translates hardcopy images into digital data that can be stored, processed, and retrieved by a computer.

Scanning Area: Full picture area scanned by a camera. This area is reduced somewhat during subsequent stages of signal processing and transmission.

Scanning Velocity: The speed at which the laser pickup head travels along the spiral track of a disc.

Scanning: Process of moving an electron beam horizontally and vertically to reproduce a television pictures. Also, process of moving a video head across a videotape to record or playback a signal.

SCART: An audio/video connector used in consumer equipment, especially in Europe. The SCART connector's 21 pins carry two audio in and out channels, in and out video channels, RGB signals, ground and some additional control signals. Only one SCART-to-SCART cable is needed to connect two VCRs or VCR to a monitor thereby avoiding the need for multiple cables.

Scatter Buying: The buying of national television advertising time one quarter at a time is known as scatter buying. Scatter buys occur after upfront and consist of all commercial inventory that was not sold during upfront. Typically scatter buys are not guaranteed and include all special programs.

Scene Chewing: An extreme, over-the-top performance that dominates the screen. "Chewing the scenery" suggests that actors are so engaged in their histrionic portrayals that furniture pieces and backdrops are left with big dental impressions.

Scene Select: A method of loading Quantel equipment that is not dependent on time code. The material is only recorded when the associated VTR is locked in play, therefore shuttling and jogging is not recorded. The material loaded in this manner can then be reordered, allowing editing during the loading process. This makes the loading time part of the edit process and not time wasted before editing can start.

Scene: A script designated locale in single-camera

production which typically constitutes a basic setting for a series of shots. Often scene is used to mean several shots, which is more to do with the word's origin in theater. It is sometimes clearer to say "sequence" for several shots, so as not to confuse the filmic and theatrical meanings of the word.

Scenery: Backgrounds and related set pieces within a setting.

Scenic Designer: An individual who carefully studies the script of a production and then works with the producer and possibly the director in devising appropriate sets.

Schedule: A list of the advertiser's commercials, time of day and dates, and the television station's or network's programs.

Schufftan Process: A technique that combines mirror shots and model shots to create a composite image. It was invented by the UFA cinematographer Eugen Schüfftan (later Schuftan) and was first used on a large scale by Fritz Lang to create the futuristic vistas of Metropolis in 1926.

Science-fiction Film: A film genre characterized by plot and action involving scientific fantasy. Also called sci-fi film.

SCMS (Serial Copy Management System): Used by DAT, MiniDisc, and other digital recording systems to control copying and limit the number of copies that can be made from copies.

Scoop: A floodlight, often used as a fill light, which has a deep, diffuse, and generally elliptical reflector.

Scopitones: Produced in the 1960s, short films of performances by popular musicians presented on coin-operated machines akin to jukeboxes.

Score: The musical component of a movie's soundtrack. Many scores are written specifically for movies by composers.

Scouting: Searching for a suitable location for doing one or more on-location scenes in a production.

Scramble: To encode an electronic signal so that it cannot be received by normal audio and video equipment. Commonly used for cable television and satellite systems to prevent subscribers from receiving specific channels without paying a subscription fee.

Scratch Mix: A mix with little correction of the sound that is usually done before the final mix in order to screen the film with all the sounds in place, to determine if there are any changes to be made. Typically this is not done on lower budget productions, as the added cost would be self-defeating.

Scratch Test: A scratch test is done before shooting, by running either a foot or two of the beginning of a roll of film, or a dummy roll of film, and checking for scratches, to insure that neither the camera nor the magazines are scratching the film.

Scratch Track: A sync recording made under conditions that make the sound useless, except for reference to the sound editor or to the actors for dubbing.

Scratch: Damage to a film in the form of a long gouge of either the emulsion or the base. A scratch on the emulsion is pretty much unfixable, since part of the image itself is missing. A scratch on the base can be alleviated with Wet Gate printing. Scratches on the workprint don't really matter at all, since you will go back to the pristine camera original for the final print.

Screen Actors Guild (SAG): Trade union for motion picture and television actors.

Screen Cards: Method used by some writers to outline their script by describing each scene on an index card, then arranging and rearranging them to work out the story structure.

Screen Direction: From the camera's perspective, the direction a character is looking and/or an object is moving in a shot.

Screen Story: Credit given to a writer who has written a screenplay based on another writer's work but has used the other writer's work only as a springboard, a characterization, an incident, or some equally limited contribution, creating a story that is substantially new and different from the other writer's work.

Screen Test: Form of audition in which an actor performs a particular role on camera, not necessarily with the correct make-up or on the set.

Screen Time: The time covered by a film's story, as opposed to its running time.

Screening: Exhibition of a movie, typically at a cinema.

Screenplay: A script written to be produced as a movie, normally between 90-120 pages. Generally speaking, a written description of a program, wherein the action and dialogue are described scene-by-scene. (Terms used to describe different types of scripts vary considerably within the television and film industries.)

Screenwriter: A writer who either adapts an existing work for production as a movie, or creates a new screenplay.

Screwball Comedy: A type of Hollywood comic film characterized by zany characters, incongruous situations, and fast-breaking events.

Scrim: A translucent sheet of material used to soften or diffuse light on a shooting set.

Script Editing: A process whereby a script is reviewed and changed, based on input from various sources such as the director or producer. Writers who specialize in script editing are called "script doctors," and are frequently uncredited.

Script Supervisor: Individual who tracks which parts have been filmed, how the filmed scenes deviated from the script; they also make continuity notes, creating a lined script.

Script: A general term for a written work detailing story, setting, and dialogue. A script may take the form of a screenplay, shooting script, lined script, continuity script, or a spec script. A script is often sold for a particular price, which is increased to a second price if the script is produced as a movie. For example, a sale may be described as "$100,000 against $250,000." In this case, the writer is paid $100,000 up front, and another $150,000 when the movie is produced.

Scrub Wheel: A mechanical control for scrubbing film or magnetic tape.

Scrub: Replay of audio tracks at a speed and pitch corresponding to jog speed, as heard with analogue audio tape "scrubbing" past an audio head. This may be provided on a digital system recording on disks to help set up cues.

SCSI (Small Computer Systems Interface): An electronic interface and command set for attaching and controlling internal or external peripherals, such as a DVD-ROM drive, to a computer. The command set of SCSI was extended for DVD-ROM devices by the SFF 8090 specification.

SDDI (Serial Digital Data Interface): A digital video interconnect designed for serial digital information to be carried over a standard SDI connection.

SDDS (Sony Dynamic Digital Sound System): A film sound format which encodes eight tracks of digital audio outside of the sprocket holes on both edges of a film print.

SDI (Serial Digital Interface): As specified in SMPTE Standard 259M.

SDMI (Secure Digital Music Initiative): Efforts and specifications for protecting digital music.

SDTI (Serial Data Transport Interface): Defined by SMPTE 305M.

SDTV (Standard Definition Television): Refers generally to digital or analog TV signals and equipment that provides picture quality roughly the same as NTSC; that is, having about 500 scan lines.

Seamless Playback: A feature of DVD-Video where a program can jump from place to place on the disc without any interruption of the video. Allows different versions of a program to be put on a single disc by sharing common parts.

Search: A mode in a video editor used to locate a specific video frame. It is also a mode in a VCR that switches a VCR into a shuttle mode so that tape can move forwards or backwards at varying speeds so that needed segments can be located and cued.

SECAM (Sequential Couleur A'memorie): The video standard used in some European and surrounding countries. In countries using the SECAM standard, most video production is done using PAL and converted to SECAM prior to transmission.

Second Assistant Camera: An assistant to the assistant cameraman.

Second Assistant Director: Assistant to the assistant director whose duties include overseeing the movements of the cast, and preparing call sheets.

Second Audio Program (SAP): A means of providing a second audio channel, for such purposes as stereophonic sound or bilingual audio tracks, on a TV broadcast channel.

Second Second Assistant Director: Assistant to the second assistant director; responsible for (among other things) directing the movements of extras.

Second Sticks!: If the clapper on the slate was not visible when the shot was being marked the camera person might call out "second sticks!" to tell the person with the slate to mark it a second time.

Second Unit: A photographic team that shoots scenes which do not involve the principal cast, such as stunts, car chases, or establishing shots.

Sector Information: Header field providing the sector number.

Sector Number: A number that uniquely identifies the physical sector on a disc.

Sector: A logical or physical group of bytes recorded on the disc-the smallest addressable unit. A DVD sector contains 38,688 bits of channel data and 2048 bytes of user data.

Seek Time: The time it takes for the head in a drive to move to a data track.

SEG (Special Effects Generator): Device designed to generate special effects. The simplest devices process a single video signal, change its color, generate sepia tones, invert the picture to a negative, posterize the image and fade or break up the image into various patterns. More sophisticated equipment uses several video sources, computer-generated graphics and sophisticated animation with digital effects.

Segment Producer: A producer who takes charge of a specific production segment, generally a segment of a newscast.

Segment Replace: The process of replacing a section of video and/or audio with different footage, irrespective of duration, with its in and out transitions re-made. This can be easily done when using true random access storage but will be more complex with linear operation and may require all subsequent material to be copied or re-edited.

Segment Rundown: A listing of the basic elements in a production in the order in which they will appear.

Segment: A piece of video and/or audio between two edit events.

Segmented Video: A videotape format in which each complete video frame is recorded through the action of two or more video heads. Most associated with 2-inch quad, the first, widely-used videotape format.

Segue: A transition from one sound to another.

Selectavision: Trademark of Thomson Electronics (which acquired Radio Corporation of America; RCA), refers to certain VCR's and a now obsolete playback only video disk system that RCA marketed.

Selective Focus: Adjusting camera focus to emphasize desired subject(s) in a shot. Selected area maintains clarity, image sharpness while remainder of image blurs. Useful for directing viewer's attention.

Selective Sound: A soundtrack that selectively includes or deletes specific sounds.

Selective Use: In the study of television stars, a use of selected parts of the star's polysemy in a particular role.

Selects: Sometimes it is useful to separate out all the shots you are going to use before beginning to edit. These are known as selects.

Self-reflexivity: A program which refers back to itself or similar programs. In a genre's evolutionary pattern, the stage during which the genre turns inward and uses its own conventions for subject matter, often in the form of a parody.

Semi-dramatic Format: A production containing dramatic scenes with actors.

Semiology: A theory of film criticism which views cinema as a language or linguistic system that conveys meaning via signs or symbolic codes. Also called Semiotics.

Semi-scripted Show: In contrast to a fully scripted show which contains all the elements and written dialogue of a show, a semi-scripted show contains only the beginning, the end, and a basic outline of the show.

Senior Spot: A spotlight with 5000 Watts of illuminating power. Also referred to as a Fiver.

Sense Memory: Technique of method acting style in which the actor draws upon memories of physical sensations of an emotional event in order to generate emotional memory.

Sensitivity: An indication of recording or playback efficiency as might be measure of a microphone or audio tape recorder.

Separation of Rights: Rights given to the creator of original written material. Includes publication, audio, live stage, live dramatic tape, live television, radio, and writer sequels, and remakes.

Separator Pages: Pages that print between job outputs to provide ease of separation are known as separator pages.

Sepia Tone: A process used in photography to generate a brownish tone in pictures giving them an "antique" appearance. The same idea has been electronically adapted for video production where a black and white image can be colored in sepia.

Sequel: A movie that presents the continuation of characters and/or events of a previously filmed movie. See also series, serial, contrast with prequel.

Sequence Shot: A long take that extends for an entire scene or sequence. It is composed of only one shot with no editing. The French term plan sequence is sometimes used because critical thinking about sequence shots owes much to French critical writing, particularly André Bazin's work on Renoir and the 1940s work of Orson Welles and William Wyler.

Sequencer: The hardware or software based brain of a MIDI studio. It receives, stores and plays back MIDI information in a desired sequence.

Sequential Scanning: Scanning lines are transmitted or reproduced in a straight numerical sequence, rather than in interlaced, odd- or even-numbered fields.

Series: A sequence of films with continuing characters or themes, but with little other interdependence, especially with respect to plot or significant character development. Until the advent of television series, there were various film series such as The Thin Man and Blondie that were started with the intention of making more than one. In subsequent years, the term would apply to features such as Star Trek: The Motion Picture or Dr. No that made more than one sequel. In modern times, the term "franchise" has been used to describe to the practice of creating a movie and product-marketing package which is contingent on commercial success. Batman and its sequels are typically used as an example of a franchise.

Serration Pulses: These are pulses that occur during the vertical sync interval of NTSC, PAL, and SECAM, at twice the normal horizontal scan rate. The reason these exist was to ensure correct 2: 1 interlacing in early televisions and eliminate DC offset buildup.

Server: System that provides large-capacity audio and video storage for video-on-demand retrieval, transmission, post production, news, etc. Most professional servers use digital disk storage.

Services Contract: Contract that a writer-client signs with an agency for representation in order to receive writing assignments.

Servo Lock: In a VTR, to lock (or synchronize) the operation of the servomechanisms to a reference sync signal.

Servo Zoom: A lens which uses a motor-driven mechanism to alter focal length.

Servo: An electronic circuit used to control the speed of a motor which drives a videotape recorder head assembly drum, which must be controlled with great precision.

SESAC: See Society of European Stage Actors and Composers.

Set Designer: Individual responsible for translating a production designer's vision of the movie's environment into a set which can be used for filming.

Set Dressing: Items of decoration which are not designated in the script or by the director as part of specific action.

Set Up: Term describing both the function of the first act in posing of the problem which the story will try to resolve, and in a more general way, the process of laying the groundwork for a dramatic or comic situation which will later be complicated, and then resolved or paid off. Set up is also used to describe the positioning of the camera and lights for a specific shot.

Set: An environment used for filming. When used in contrast to location, it refers to one artificially constructed. A set typically is not a complete or accurate replica of the environment as defined by the script, but is carefully constructed to make filming easier but still appear natural when viewed from the camera angle.

Set-top Box (STB): Device that converts and displays data from analog cable, digital cable, or digital broadcast television to a standard frequency (channel number) for display on a standard analog television set, or a box that receives off-air DTV signal for display on a DTV monitor.

SFF 8090: Specification number 8090 of the Small Form Factor Committee, an ad hoc group formed to promptly address disk industry needs and to develop recommendations to be passed on to standards organizations. SFF 8090 (also known as the Mt. Fuji specification), defines a command set for CD-ROM- and DVD-ROM-type devices, including implementation notes for ATAPI and SCSI.

Shader: The video engineer responsible for maintaining the luminance and color quality of a camera.

Shading: To adjust the chroma and video levels on a video camera. Often, an on-going process that takes place throughout the broadcast or taping of a production.

Shadow Mask: A perforated metal plate within a color TV picture tube that controls which colored dots or stripes the electron beams will activate. The shadow mask keeps the specific electron beams from hitting the phosphors of the wrong color.

Shallow Focus: A small depth of field, with just one plane (foreground, middle-ground, or background) in focus.

Shallow Space Blocking: A type of blocking associated

with multiple-camera, studio set productions, where, due to the shallow sets, the actors mostly move side-to-side, rather than up-and-back.

Share: In the context of TV ratings, the percentage of homes with turned-on television sets that are tuned to a specific program. Usually used in conjunction with ratings.

Sharpie: A permanent felt-tipped marker useful for labeling the cans of exposed rolls out on a shoot and in the editing room for labeling the leader. Sharpie is a brand-name of the most common of these markers.

Sharpness: Apparent image resolution or clarity of an image.

SHF (Super High Frequency): The frequencies ranging from 3 GHz to 30 GHz. These include communications satellite signals and most microwave transmissions.

Shield Law: A law that exists in many states that keeps a news reporter from having to reveal confidential sources of information.

Shield: Commonly, a woven braid around an audio line that reduces or eliminates outside interference.

Shift Register: An electronic memory bank associated with CCDs which holds a charge before being read out by the system's scanning process.

Shimmering: Flickering halos or pinpoints of light caused by stray light rays from imperfect rear projection screen layers, which contain lens elements. Also accentuated flicker of thin picture details moving up or down at certain speeds that interact with the scan rate.

Shiny Boards: A grip reflector used for re-aiming sunlight to provide a key or fill light.

Shock Cut: A jarring transition between two actions occurring at different times or places. Also called a smash cut.

Shock Mount: A rubber or spring-based microphone holder which reduces or eliminates the transfer of sound or vibration to the microphone.

Shock Value: An element in news value based on the amount of amazement, surprise, horror or scandal represented by a story.

Shoot and Protect: A term which relates to the need to keep the full HDTV 16:9 aspect ratio free of extraneous subject matter, even if it is reproduced in a 4X3 ratio. In this case action should be confined to the smaller area (the shoot range), but the larger area (the protect range) must be free of microphone booms, lights and other distracting elements.

Shooting Ratio: The amount of film footage shot compared to the length of the film's final cut. For instance a 5 minute film for which you shot 30 minutes of footage would have a shooting ratio of 6 to 1.

Shooting Schedule: Production schedule for shooting a film with the scenes from a script grouped together and ordered with production considerations in mind.

Shooting Script: A script from which a movie is made which contains that includes scene numbers, camera angles, inserts, and certain directors/cinematographers input.

Short: Movie that is shorter than 60 minutes.

Shortends: The unexposed remainder of a roll of film in a magazine that is clipped and placed back into a can for use later. Unlike recans a shortend is something less than 400 feet.

Shortform: Short-length programs, not longform. Editing shortform programs can be very complex as they include commercials, promotions, television station IDs, bumpers and stings. Shortform editing systems do not have to store and handle large amounts of material but make heavy use of effects and layering. See also Longform.

Shot Box: A control that can electronically present various zoom speeds and focal lengths for a zoom lens. Preset positions are activated by push buttons.

Shot Composition: The arrangement of key elements within the frame.

Shot Sheet: A listing of all shots recorded on a videotape along with times and descriptive comments. Also a list of shots required by a camera operator.

Shot: A single, continuous run of the camera. The images recorded on a strip of exposed film from the time the camera starts until the time it stops.

Shot-by-shot Analysis: Close and thorough study of the separate shots that make up a scene, sequence, or film. Also called Shot Analysis.

Shot-by-shot Script Style: As opposed to the master scene script style, which outlines only general scenes, the shot-by-shot script style lists each individual shot.

Shotgun Microphone: A highly directional microphone, usually with a long, tubular body; used by the production sound mixer on location or on the set for film and television productions.

Shoulder Brace: A support for a video camera that rests on both the shoulder and the body.

Showcard: A white artists' cardboard which is used as a reflector or for making other special rigs. It is easily cut and formed.

Showscan: A film process using 70mm film (65mm in the camera) shot and projected at 60 frames per-second (rather than 24). The process results in a much greater fidelity than traditional motion picture methods.

Shutter Speed: The length of time for which the shutter stays open. The higher the shutter speed is, the more clearly a moving object can be shot.

Shutter Speed: The length of time that a single frame is exposed for. Slower shutter speeds allow more light to enter the camera, but allow more motion blur.

Shutter: The shutter in film cameras is a circular disc that changes in shape to increase or decrease the exposure of the film. Besides the aperture and film it is one of the most important parts of a camera. It was the camera part that took longest to develop when cinema was first introduced.

Shuttle: Where playback speed is governed by the position of a control. Shuttle speeds typically range +/- 50 x play. Shuttle is not used in random access editing as all material can be instantly accessed.

SI (Système International): A complete system of standardized units and prefixes for fundamental quantities of length, time, volume, mass, and so on.

Siamese: A splitter that divides a power line into two parts.

Sibilance: An exaggerated hissing in voice patterns.

Side Lighting: Light source to one side of the subject, leaving a shadow on the opposite side.

Side Panels: Used with imaging a standard 4:3 picture on a wide-screen 16: 9 aspect ratio television screen, typically with black bars (side panels) on each side. Used to maintain the original aspect ratio of the source material.

Sider: A device which cuts the light from the side of a lighting unit, usually a flag or a cutter.

SIF (Standard Input Format): This video format was developed to allow the storage and transmission of digital video. The 625/50 SIF format has a resolution of 352 x 288 active pixels and a refresh rate of 25 frames per second. The 525/60 SIF format has a resolution of 352 x 240 active pixels and a refresh rate of 30 frames per second. Not that MPEG-1 allows resolutions up to 4095 x 4095 active pixels; however, there is a constrained subset" of parameters defined as SIF. The computer industry, which uses square pixels has defined SIF to be 320 x 240 active pixels, with a refresh rate of whatever the computer is capable of supporting.

Sight Gag: A visual joke; a piece of non-verbal comic

business in a film.

Sight Line: An imaginary line that is drawn between a subject and the object that he/she is looking at.

Sign: In semiotics, the smallest unit of meaning: composed of a signifier and its signified.

Signal: The form of variation with time of a wave whereby information is conveyed in some form whether it is acoustic or electronic.

Signifier: The physical aspect of a sign, such as ink on a page, chalk on a chalkboard, a blinking light, light emanating from a TV screen, etc.

Signs of Character: The various signifiers—viewer foreknowledge, character name, appearance, objective correlatives, dialogue, lighting, videography or cinematography, and action—that communicate the character to the viewer.

Signs of Performance: The actor's facial, gestural, corporeal and vocal signifiers that contribute to the development of character.

Silent Camera: This term is often a little confusing because it does not mean a camera that is itself silent, and therefore usable for sync sound, but it means a noisy, un-silent camera, usable only for shooting silent, M.O.S. scenes.

Silent Film: A film that has no synchronized soundtrack and no spoken dialogue. It was a form predominate in film until the late 1920's when practical synchronized soundtrack technology was developed and its use became popular.

Silent Radio: Silent radio is a service that feeds data that is often seen in hotels and nightclubs. It's usually a large red sigh that shows current news, events, scores, etc. It is present on NTSC lines 10-11 and 273-274, and uses encoding similar to EIA-608.

Silent Speed: 18 frames per second. A slightly archaic

notion left over from the time when 16mm was used exclusively for home movies. It is not always that easy to find a projector that will project at 18 frames per second and so films shot at silent speed will often be speeded up slightly, whether the filmmaker intended this of not.

Silk: A lighting diffusion or reflective material, formerly real silk.

Simmons Market Research Bureau (SMRB): SMRB is a research company that publishes an annual syndicated report measuring the cumulative viewing of cable networks, product usage, psychographics, magazine readership, etc. SMRB's software package is called CHOICES II.

Simple Profile (SP): A subset of the syntax of the MPEG-2 video standard designed for simple and inexpensive applications such as software. SP does not allow B pictures. See also Profile.

Simulate: To test the function of a DVD disc in the authoring system, without actually formatting an image.

Simulcast: The broadcast of the same program simultaneously over two or more different systems or channels. An accelerated amount of simulcasting of both analog and digital programming will be required by FCC rules, during the DTV transition period.

Simultaneous Contrast: The effect of a surrounding tone or color on a specific color or tonal value. Certain color combinations or brightness values, when placed in proximity, interact and affect each other.

Single Perf: 16mm film with a row of perforations along one edge. On the film can this will be indicated by 1R appearing on the label.

Single Reel: In 35mm a reel is 1,000 feet of film (or usually a little less).

Single System: Single System refers to recording, editing or projecting sound and picture together on the same piece of film. Cameras used for TV news would record the sound on a magnetic stripe as well as photograph the picture. Also super-8 sound. Single system has some distinct editorial disadvantages, hence the more common use of Double System for shooting and editing.

Single: A shot with only one subject in the frame.

Single-frame Cinematography: Shooting film one frame at a time to speed up normal motion, to make lifeless objects appear to move, or to do time-lapse photography. Also called Single-framing.

Single-frame Recorder: A VCR capable of recording one video frame at a time. Commonly used in doing animation.

Single-stripe: Magnetic film that contains a single audio track, which is coated with oxide.

Situation Comedy: A comedy in which humor is derived from people being placed in uncomfortable, embarrassing, or unfamiliar situations.

Skew Control: A playback control on a videotape machine which adjusts videotape tension and the length of the video tracks read from the videotape. Improper skew adjustment results in various picture aberrations.

Skip Frame: An optical printing effect which eliminates selected frames of the original scene to speed up the action.

Skylight: (1A) or haze (UV) filter mounted at the front of a camcorder lens; clear glass absorbs ultraviolet light. Also excellent as a constant lens protector.

Slander: Uttering false statements deemed harmful to another person's character or reputation.

Slant Track: Helical scan videotape recorder.

Slapstick Comedy: A comedy in which the humor is derived from physical interactions, often involving exaggerated but ultimately harmless violence directed towards individuals. Named after a device used by circus clowns—two boards which slap together loudly when swatted on someone's backside.

Slate: A board upon which key information about a shot is displayed (i.e., scene and take numbers, title of the show, whether it is day or night, sync, or MOS). This board is held in view of the camera either at the head or tail of a shot to identify it to the lab and to the editor. If it appears at the tail of a shot, it will be held upside-down.

Slave: An audio tape or videotape transport, projector or mag film dubber whose movements follow the movement of a single master transport. Accomplished electronically by using SMPTE time code numbers or mechanically by motor linkage of sprocketed machines.

Sliced VBI Data: A technique where a VBI decoder samples the VBI data (such as teletext and captioning data), locks to the timing information, and converts it to binary 0's and 1's. DC offsets, amplitude variations, and ghosting must be compensated for by the VBI decoder to accurately recover the data.

Slide Fader: See Linear Fader.

Slip-cuing: Method of cueing records which involves moving the record forwards and backwards until the precise starting point of a selection is found. Allows for the instantaneous start of a selection.

Slop Print: An untimed black and white dupe print of the workprint, used for projection in a sound mix. A slop print is used because splices can jump and cause the film to go out of sync, and a slop print will have no splices.

Slot: As in an actor's availability, principal actors have availabilities between previous commitments that are described as "slots."

Slow Motion: Playback of a videotape at a speed slower than the speed at which it was recorded.

Slow-speed Film: A fine-grain stock that registers a wide range of subtle gradations between black and white. Requires the greatest amount of light. Also called Slow Film.

Slug Line: A header appearing in a script before each scene or shot detailing the location, date, and time that the following action is intended to occur in.

Slug: A strip of blank leader or image-bearing film used as leader.

Slung Microphone: A microphone that is suspended above performers by wire, cord, or rope.

Smash Cut: See Shock Cut.

Smear: A trailing image associated with motion caused by any one of several video problems. Normally associated with Vidicon, tube-type cameras when used under low light levels.

SMPTE 120M: NTSC video specification for the United States. See RS-170A and BT.470.

SMPTE 125M: 720 x 480 pro-video interlaced standard (29.97 Hz). Covers the digital representation and the digital parallel interface. Also see BT.601 and BT.656.

SMPTE 12M: Defines the longitudinal (LTC) and vertical interval (VITC) time code for NTSC and PAL video systems. LTC requires an entire field time to store time code information, using a separate track. VITC uses one scan line each field during the vertical blanking interval.

SMPTE 240M: 1920 x 1035 pro video interlaced standard (29.97 or 30 Hz). Covers the analog RGB and YpbPr representation. The digital parallel interface is defined by SMPTE 260M. The digital serial interface is defined by SMPTE 292M.

SMPTE 244M: 768 x 486 pro-video interlaced standard (29.97 Hz). Covers the digital representation (composite NTSC video sampled at 4x Fsc) and the digital parallel interface. The digital serial interface is defined by SMPTE 259M.

SMPTE 260M: Digital representation and parallel interface for SMPTE 240M video.

SMPTE 266M: Defines the digital vertical interval time code (DVITC). Also see BT.1366.

SMPTE 267M: 960 x 480 pro-video interlaced standard (29.97 Hz). Covers the digital representation and the digital parallel interface. Also see BT.601 and BT.1302.

SMPTE 272M: Formatting AES/EBU digital audio and auxiliary data into the digital blanking intervals. Also see BT.1305.

SMPTE 274M: 1920 x 1080 pro-video interlaced and progressive standards. Covers the digital representation, the analog RGB and YpbPr interfaces, and the digital parallel interface. The digital serial interface is defined by SMPTE 292M.

SMPTE 276M: Transmission of AES/EBU digital audio and auxiliary data over coaxial cable.

SMPTE 291M: Ancillary data packet and space formatting for pro-video digital interfaces. Also see BT.1364.

SMPTE 292M: 1.485 Gbps pro-video HDTV serial interfaces.

SMPTE 293M: 720 x 480 pro-video progressive standards (59.94 Hz). Covers the digital representation, the analog RGB and YPbPr interfaces, and the digital parallel interface. The digital serial interface is defined by SMPTE 294M. Also see BT.1358 and BT.1362.

SMPTE 294M: Pro-video serial digital interface for SMPTE 293M.

SMPTE 296M: 1280 x 720 pro-video progressive standards. Covers the digital representation and the analog RGB and YPbPr interfaces. The digital parallel interface uses SMPTE 274M. The digital serial interface is defined by SMPTE 292M.

SMPTE 299M: 24-bit digital audio format for pro-video HDTV serial interfaces. Also see BT.1365.

SMPTE 305M: Serial data transport interface (SDTI). This is a 270 or 360 Mbps serial interface based on BT.656 that can be used to transfer almost any type of digital data, including MPEG-2 program streams, MPEG-2 transport streams, DV bit streams, etc. You cannot exchange material between devices that use different data types. Material that is created in one data type can only be transported to other devices that support the same data type. There are separate map documents that format each data type into the 305M transport.

SMPTE 308M: MPEG-2 4**:** 2**:** 2 profile at high level.

SMPTE 314M: Data structure for DV-based audio, data and compressed video at 25 and 50 Mbps. Also see IEC 61834.

SMPTE 322M: Data stream format for the exchange of DV-based audio, data and compressed video over a Serial Data Transport Interface (SDTI or SMPTE 305M).

SMPTE 344M: Defines a 540 Mbps serial digital interface for pro-video applications.

SMPTE 348M: High data-rate serial data transport interface (HD-SDTI). This is a 1.485 Gbps serial interface based on SMPTE 292M that can be used to transfer almost any type of digital data, including MPEG-2 program streams, MPEG-2 transport streams, DV bit streams, etc. You cannot exchange material between devices that use different data types. Material that is created in one data type can only be transported to other devices that support the same data type. There are separate map documents that format each data type into the 348M transport.

SMPTE Leader: A leader placed at the head of release prints containing information for the projectionist and featuring numbers which are black on a medium density background. These numbers count down from 8 to 2 at 24 frame intervals ending at the first frame of the "2" followed by 47 frames of black.

SMPTE Leader: Another term for Academy Leader.

SMPTE RP160: Analog RGB and YPbPr video interface specification for pro-video HDTV systems.

SMPTE/EBU Time Code: A digital electronic signal recorded on a video or audio tape that provides a precise time-based numbering system together with other information. Also called Address Track. The name designates a standard agreed upon by the Society of Motion Picture and Television Engineers and the European Broadcast Union.

SMPTE: See Society of Motion Picture and Television Engineers.

SMPTE-VITC: SMPTE's vertical interval time code (VITC) format standard. The term VITC, used alone, usually refers to SMPTE-VITC.

SMRB: See Simmons Market Research Bureau.

SMTPE 253M: Analog RGB video interface specification for pro-video SDTV systems.

SMTPE 259M: Pro-video serial digital interface for SMPTE 244M.

Snake: A multi-channel audio cable intended for use with microphone level signals and/or line level signals.

Snap Zoom: A very rapid zoom from a wide-angle into a close-up of some person or object intended to achieve a dramatic effect.

Sneak Preview: An unannounced screening of a movie before the premiere, often used to gauge.

Snoot: Open-ended cylindrical funnel mounted on a light source to project a narrow, concentrated circle of illumination.

Snow: A general term used to describe interference in a video image. It manifests as random colored or black and white dots.

SNV: See Satellite News Vehicle.

Social Actor: "Real" people as used in nonfiction television programs; people "performing" according to social codes of behavior in order to represent themselves to others.

Society of European Stage Actors and Composers

(SESAC): An agency that licenses copyrighted music.

Society of Motion Picture and Television Engineers (SMPTE): Technical society devoted to advancing the theory and application of motion-imaging technology including film, television, video, computer imaging, and telecommunications.

SOF (Sound-on-film): A film that includes a soundtrack.

Soft Contrast Filter: A filter that reduces the contrast ratio of a scene, typically used to bring it within the desired 30:1 video ratio.

Soft Focus: A glamorizing technique to create a blurred or hazy effect, achieved by shooting slightly out of focus or through gauze, filter, special lens, or Vaseline.

Soft Light: A type of light with a built-in surface to act as a bounce card, providing soft, indirect light on the subject.

Soft Matte: Refers to widescreen movies that were actually filmed wide-angle using the entire 4:3 aspect ratio film frame and where the projectionist adjusts mattes (aperture plates) to hide the top and bottom edges leaving the middle as a widescreen view. Sometimes what will become a widescreen film is shot in the 4:3 aspect ratio so that when committed to video, the entire width of the picture can be included and also the entire TV screen would be filled.

Soft News: News stories that examine the personal, such as gossip, scandal, murder, mayhem, and "human interest" stories.

Soft Wipe: A video wipe in which the demarcation line between the two sources blends into a soft line.

Soft-copy Prompter: A camera prompting device that uses a video display, as opposed to a hard-copy prompter that uses paper.

Solarization: Special effect in which the lightest and darkest values of a picture are made dark while the middle tones become light.

Son: The metal disc produced from a mother disc in the replication process. Fathers or sons are used in molds to stamp discs.

Sound Bite: Any recorded video or audio-only segment salvageable for use in edited program—usually a highlight phrase or event. Common component of broadcast news.

Sound Blanket: Basically just a quilted mover's blanket. Often it is thrown over the camera (and the camera operator) to cut down on camera noise, as a sort of improvised Barney.

Sound Bridge: Sound which continues across two shots that depict action in different times or places, thus providing an aural transition between the two scenes.

Sound Crew: Group of crewmembers directly involved with creating of a movie's soundtrack: sound designer, sound editor, sound effects, sound mixer, sound recordist, boom operator, re-recording mixer, music supervisor, and foley artist.

Sound Designer: The conceptual chief of a movie's soundtrack, responsible for designing and creating the audio component of a movie.

Sound Dubbing: The completion of the sound edit. This can take place either before or after the on-line: or be part of it. Sound dubbing may involve adding music and sound effects, tidying sound edits and equalizing different pieces of audio, all with the aim of producing a final soundtrack.

Sound Editor: Member of the sound crew who performs editing on the soundtrack. See also dialog editor.

Sound Effects: Any sound in a film other than dialogue, narration, or music.

Sound Fill: See Filler.

Sound Reader: A playback head for reading mag stock, mounted on a bracket that snaps onto a synchronizer. It is plugged into the squawk box.

Sound Recordist: Member of the sound crew responsible for operating the audio recording equipment on a set.

Sound Speed: 24 frames per second. The normal speed for filming and projecting.

Sound Stage: A large room designed for the filming or videotaping of programs. Sets are arranged on the stage in a variety of ways, depending mostly upon the presence/absence of a studio audience.

Soundies: Produced in the 1940s, short films of performances by popular musicians presented on coin-operated machines akin to jukeboxes.

Soundtrack: The optical or magnetic strip at the edge of the film which carries the sound. Also, any length of film carrying only sound. Popularly, it refers to a collection of songs which are heard during the movie, often sold as an album.

Source Machine: In videotape editing, the machine(s) that contain the original footage that will be edited onto the edited master.

SP (Standard Play): Fastest tape speed of a VHS VCR, accommodating two-hour recordings.

SP: See Simple Profile.

SP@ML (Simple Profile at Main Level): The simplest MPEG-2 format used by DVD. Most discs use MP@ML. SP does not allow B pictures.

Space: The reflective area of a writable optical disc. Equivalent to a land.

Spacer: A metal cylinder with a flat plate at one end and a hole through the center, used between reels on the spindle of a rewind to space out the reels the same distance as the gangs of a synchronizer. Although it is a little shorter, in a pinch you can use cores as spacers.

Sparklies: Flecks of snow in a satellite reception. See Impulse Noise.

Spatial Compression: A compression technique that reduces the information needed to display video frames by removing or reordering information regarding the field of colored pixels.

Spatial Resolution: The clarity of a single image or the measure of detail in an image. See resolution.

Spatial: Relating to space, usually two-dimensional. Video can be defined by its spatial characteristics (information from the horizontal plane and vertical plane) and its temporal characteristics (information at different instances in time).

Spatially Adjacent: Refers to scan lines immediately juxtaposed on the screen. The term is needed when discussing interlaced video where juxtaposed lines are (for NTSC) 1/60 second apart in terms of when they were transmitted, received, and "drawn.."

Speaking Role: A role is one in which the character speaks scripted dialogue in contrast to a "non-speaking role" where the character is specifically mentioned in the script but who doesn't have any lines of dialogue in the finished film.

Spec Script: A script written before any agreement has been entered into ("on spec" or speculation), in hopes of selling the script to the highest bidder once it has been completed.

Special Effects Generator (SEG): Video signal processor with vast, but varying, image manipulation capabilities involving patterns and placement as well as color and texture: mixing, multiplying, shrinking, strobing, wiping, dissolving, flipping, colorizing, etc.

Special Effects Supervisor: Chief of a production's special effects crew.

Special Effects: Shots which are unobtainable by straightforward motion picture filming techniques and which require special models, matting, multiple exposure, etc. The term also applies to most pyrotechnic and ballistic effects in a film.

Special Event: A program that will be run only once or a small, defined number of times.

Specifics: In sound editing, these are any effects that directly relate to the picture, where we see a thing happen and hear it too. Backgrounds, ambiance and speech are not specifics.

Spectral Highlights: Bright reflections from shiny subject matter that often cause spikes on a waveform monitor and video brightness range problems.

Spectrum Analyzer: A receiver capable of receiving, monitoring, and analyzing specific RF frequencies. Often used in SNG work.

Spectrum: Called "the invisible highway," spectrum is the collection of radio frequencies used by analog television broadcasters. Other users of spectrum include police and fire department radios, air traffic control, satellite transmissions, microwave ovens, cell phones and baby monitors.

Specular: A term used to describe highly directional, focused light. This is often perceived as a very "hard" light.

Speed of Sound: The velocity of sound in air is 770 mi/hr. This speed however, is influenced by temperature and air pressure.

Speed!: This is what the cameraperson or sound recordist will call out to acknowledge that they are rolling. It comes from the days when it took a few seconds for certain equipment to reach proper speed.

Speed: When we include all the transfer times, the effect of a shutter, if any, the effective speed of the system shall be expressed in FPS (frames per second). The base speed for a digital production camera system shall be 24 FPS.

Spherical: An optical system in which the magnifications of the vertical and horizontal dimensions of the picture are the same. See also Aspect Ratio.

Spider: Another, less commonly used term for Spreader.

Spikes: Spikes are a term that comes from theater. They are little pieces of tape placed around the legs of furniture, or the tripod legs, before they are moved, making it easy to return things to their original position.

Spill: Light that is escaping from the sides of a lighting unit, or any light that is falling where it is not wanted.

Spill-in: A spill-in is the percent or number of households in a home market viewing television stations in an outside market.

Spill-out: A spill-out is the percent or number of households in an outside market viewing television stations originating in a home market.

Spins: Announcements airing in a time period or program other than the rated barter syndicated program for a particular market.

Splice: A method of joining two pieces of film so they can be projected as one continuous piece. There are three methods: the Tape Splice (usually used for editing), the Cement Splice (used for original material), and the far less common Ultra-Sonic Splice (used for Polyester Base film).

Splicing Tape: A special type of clear tape, not interchangeable with scotch tape, used to splice film. It comes in perforated (for use with a Rivas) and unperforated (for use with a Guillotine). Transparent splicing tape is used for picture and white splicing tape for sound.

Split Edit: An edit where the audio and video start at different points, for example hearing the first few words of an interview before we see the person speaking.

Split Field Lens: A "bifocal-type" lens that fits over the front of the standard camera lens and makes it possible to divide the scene and bring near and far objects into focus at the same time.

Split Reel: A very handy reel, the two halves of which may be unscrewed and film on a core placed between. Once screwed back together (but not too tight, or it will never open) the film on a core has quickly been converted into film on a reel.

Split Screen: TV or computer screen electronically divided to show two or more images at once.

Split Sync Scrambling: Split sync is a video scrambling technique, usually used with either horizontal blanking inversion, active video inversion, or both. In split sync, the horizontal sync pulse is "split," with the second half of the pulse at +100 IRE instead of the standard -40 IRE. Depending on the scrambling mode, either the entire horizontal blanking interval is inverted about the +30 IRE axis, the active video is inverted about the +30 IRE axis, both are inverted, or neither is inverted. By splitting the horizontal sync pulse, a reference of both -40 IRE and +100 IRE is available to the descrambler.

Split-session: A job spread over two or more (post production/edit) sessions. This may be for scheduling, overruns, etc. Operating with tape, all footage and an EDL can be off-loaded from the machines in the suite and re-loaded later in the same suite, or elsewhere. Using disk-based systems the footage is stored on non-removable media so the session has to be archived to later re-load the session in its existing state. Clearly the speed and completeness of the archiving system is important to this operation. See also Archive and Rebuild.

Spool Down: Winding an unexposed 400 foot roll down onto four 100 foot daylight spools for use in a camera that will only take 100 feet of film. Spooling down can only be done in complete darkness. 42 turns on a rewind per daylight spool will divide a 400 foot roll pretty evenly. Also, it is vitally important that the film be wound all the way through once and then spooled down; otherwise the edge numbers will be on the wrong side, and not printed onto the workprint.

Spool: A flanged roll on which film is wound for general handling or projection.

Spot Meter: A type of meter for taking a Reflective Light Reading with a short telescopic sight that enables you to take a very specific reflective reading of a small, well-defined area.

Spot: A commercial announcement, typically less than five minutes in length, that is placed within or next to a program. The most common spot lengths are 30 seconds and 60 seconds. On a lensed light, the smallest beam spread.

Spotlight Effect: Video effect resulting in a circular section of the screen appearing brighter in order to highlight a particular element.

Spotlight: A studio lamp that illuminates a relatively small, specific area, or "spot."

Spots Per Week (SPW): Known as SPW, ordering spots without specifying on which days of the week the spots should run is called a spots-per-week buy. Acceptable or active days of the week may be specified and minimum and maximum activity defined for each day.

Spotting: Used in scoring and sound effects editing to identify the specific scenes or points where music cues or effects cues will take place. Usually, this will include information on length and style.

Spreader: A piece of gear consisting of three arms on a central hub attached to the bottom of a tripod to keep the legs from collapsing outwards.

Spring Lock: A round spring-loaded clamp that goes on the end of a rewind to allow several reels to turn together.

Sprite: A 2-D object drawn in the 3-D environment on a single polygon with a transparency, always facing the viewpoint of the player. All the characters in Doom were sprites, vs. real 3-D characters in Quake or GK3. Sprites are still used for some effects like fire, explosions, plants—things that have rough edges to they or need to be heavily animated.

Sprocket Holes: The same as Perf.

Sprocket: The teeth on a roller designed to engage with the perforations in film. Sometimes sprocket holes are referred to as sprockets too.

Spun: Spun glass diffusion material. See also Diffusion.

SPW: See Spots Per Week.

Squawk Box: A small amplified speaker used on an editing bench and receiving sound from the Sound Reader.

Squeeze Trick: See Vertical Squeeze Trick.

Squeezed Video: See Anamorphic.

Squib: A small explosive device, which when detonated will simulate the effect of a bullet/puncture wound or small explosion. When worn by actors, they typically include a container of blood which bursts upon detonation. See also Special Makeup Effects.

Stabilization: Removal of unwanted movement of pictures, such as film weave or camera shake. This may be achieved automatically by applying tracking data to DVE facilities.

Stabilizer: Video signal processor used primarily for tape dubbing to eliminate picture jump and jitter, maintain stability.

Stacking Ring: A circular ridge used to separate the discs during the processing to stop them sticking.

Stage Box: A distribution box with six pockets for stage plug connectors.

Stagehand: An assistant who can be assigned a wide variety of studio responsibilities including setting up and moving sets, props, and equipment.

Staging: The re-enactment of an event for the purposes of video production. In doing news, staging refers to the sometimes questionable re-enactment of a happening for the purposes of a news story. Also, a temporary structure used as a production setting.

Stairstepping: The jagged appearance of sloped lines or edges inherent to the reproduction of a picture using pixels and/or horizontal scan lines.

Stamping: The process of replicating optical discs by injecting liquid plastic into a mold containing a stamper (father or son). Also (inaccurately) called mastering.

Standard Billing Month: Also referred to as "broadcast month," this is the month advertising agencies frequently use for billing purposes. It is defined as a fiscal month that ends on the last Sunday of a calendar month. This month can have four or five weeks in hit, but it will always allow for a complete Monday through Sunday billing week.

Standard Input Format: See SIF.

Standard Week: The seven days from Monday through Sunday, inclusive, and in that order. Used extensively for the scheduling of commercial time, the standard week is also known as the Media Week.

Standards Converter: A device for converting signals from one broadcast standard to another.

Stand-In: Individual who has the same physical properties of a particular actor, and takes their place during the lengthy setup of a scene.

Standing Waves: A deep sound in a small room or booth from low frequency caused by long waves with short reflection patterns.

Stand-up: The feature of a television news package, in which the reporter stands before a site significant to the story to narrate it.

Stanislavsky Method: See Method Acting.

Star Filter: An optical filter that has finely etched criss-crossing lines on the surface that creates fingers of light around bright lights and spectral reflections.

Star Image: A representation of an actor that is fabricated through the media texts of promotion, publicity, television programs, and criticism.

Star System: A system developed in the early days of Hollywood to market movies based on the appeal of popular actors and actresses, "movie stars," who were under contract with commercial motion picture studios to play leading roles in their productions.

Starsight: An electronic program guide that you subscribe to. It allows you to sort the guide by the order of preference and delete television stations you never watch. It's a national service that is regionalized. The decoders in Houston only download data for Houston. Move to Dallas and you only get Dallas.

Static Composition: Elements and guidelines of composition related to still images. See also Dynamic Composition.

Status Indicators: The collection of various lights, patterns and alphanumeric characters visible in and around a video camera viewfinder that show the operating condition of the camera, recorder and battery.

STB: See Set-top Box.

Steadicam: A brand-name for a body camera mounts which uses a system of counterbalanced springs to keep a camera reasonably steady, even when the camera operator is walking or running. The steadicam allows a camera to be mounted on a person but it absorbs most of persons shocks or movements. This allows for fluent hand-held shots. Camera movements done this way are very similar to those done on a dolly. The steadicam in film is great because it cuts down the shot setup time normally required.

Steenbeck: A popular brand of flatbed. The word is sometimes used interchangeably with flatbed.

Step Outline: Method used by some writers to outline their story by numbering the major scenes and the order in which they occur.

Step Printer: A printer in which each frame of the negative and raw stock is television stationary at the time of exposure.

Stereo Microphone: A microphone containing two sound-sensitive elements designed and arranged in such a way as to reproduce a stereo sound perspective.

Stereo Mixing: Simultaneous processing of both left and right audio channels.

Stereo Separation: The degree to which left and right stereo channel information is perceived as being distinct and separate.

Stereo Synthesizer: An electronic circuit that modifies a monophonic audio signal to simulate stereo.

Stereo: Sound emanating from two isolated sources, intended to simulate pattern of natural human hearing.

Stereotype: A conventionalized character type that is demeaning to a particular social group.

Sticks: A slang term for a camera tripod.

Still Frame Storage Unit: A digital device that stores individual video frames which can be instantly recalled by entering an address number. It eliminates the need for slides and camera cards.

Still Frame: A graphic of any kind presented as a single, static video image.

Still Photographer: Individual who photographs the action (often alongside the camera) to be used in publicizing the movie.

Still Video Camera: A camera that allows still video frames to be stored on small computer-type magnetic disks or removable solid-state devices.

Still: A photograph taken of a film scene for promotional purposes, not to be confused with a frame enlargement reproduced from actual film footage.

Stinger: A single extension cord. Most often referred to a single "hot" extension that is left lying around for occasional use.

Stock Footage: Motion picture footage from previously existing films, filed in film libraries or archives, which is incorporated into a new film, usually to provide scenic background or for stock situations and settings, such as war scenes or foreign locations. See also Compilation Film.

Stock Shot: Common footage, for example city traffic or a rainbow that is conveniently accessed as needed. Similar to a "photo file" in photography.

Stop Frame: An optical printing effect in which a single frame image is repeated in order to appear television stationary when it is projected. This may also refer to a camera technique in which only one frame at a time is exposed.

Stop Motion: A form of animation in which objects are filmed frame-by-frame and altered slightly in between each frame. See also Go Motion.

Stop Set: A cluster or group of commercials.

Stop-motion Photography: Filming real objects or live action by starting and stopping the camera, rather than by running the camera continuously, in order to create pixilation, trick-film effects, or time-lapse photography. Also called Stop-action Photography.

Stops: The number of stops represents the range of light intensity which the system can effectively portray. A stop on a "Digital Production Camera" will be equal to an F-stop on a film camera.

Story Time: The amount of time that transpires within a program's narrative. See Screen Time.

Storyboard: A pictorial outline of a film scene or sequence using drawings or photographs to illustrate how each shot should look on the screen.

STP (Surface Transfer Process): A method of producing dual-layer DVDs that sputters the reflective (aluminum) layer onto a temporary substrate of PMMA, then transfers the metalized layer to the already-molded layer 0.

STP: See Surface Transfer Process.

Straight Cut: Two shots edited together without optical effects.

Streaking: Smearing effect around objects in a picture caused by aberrations in the video system.

Streamer: A grease pencil mark on the workprint indicating either a fade or a dissolve, called so because when projected it resembles a streamer trailing across the screen.

Streaming Media: A technique for transferring digital media on the Internet, allowing the user to play media from the Internet immediately, without downloading the entire media file, eliminating the wait of downloading. The advantage for the copyholder is that the end user does not receive a storable copy of the file.

Stretch: Process that performs speed-up and slow-down of video. In Quantel editing systems this provides interpolated results unmatched by VTR-based technology.

Strike: To take down and remove scenery after a production.

Striking: The breakdown process of a camera position, location, or set.

Strip Show: A program or commercial announcement scheduled at the same time on successive days of the week, either Monday through Friday or Monday through Sunday, is known as a strip show.

Stripe: 35mm mag stock that contains a stripe of magnetic tape rather than the complete coating found on Fullcoat. Stripe mag will also have a balance stripe to prevent warping.

Striped Filter: Microscopic vertical stripes on the surface of a color CCD imaging device designed to break up the picture into two basic color components. Serves the same basic function of dichroic mirrors in three-CCD or three-tube cameras.

Striped Tape: A tape that has continuous control track and time code recorded along its length making it ready for use as a recording tape for insert editing. It is very similar to formatting a disk in that it is the preparation of a medium for use.

Stripped Syndication: A programming strategy in which syndicated shows are scheduled Monday through Friday in the same time slot.

Strobe: Special effect in which a frame is periodically held for a finite time until another frame is held.

Structuralism: Cinematic theories focusing on how certain codes or signs are structured to convey meaning in a film, a genre, or the works of a filmmaker. See also Semiology.

Structured Polysemy: The organization and emphasis/repression of meanings within television's polysemy.

Studio Crane: See Crane Shot.

Studio Floor Plan: A scale drawing of a studio area showing where scenery and props (and sometimes lights and cameras) are to be placed.

Studio Set: Three-walled, ceiling-less set erected on a sound stage; this type of set is usually shallow, normally wider than deep, and rectangular rather than square.

Studio: A company that makes movies. Larger studios (such as the majors) have extensive in-house soundstages where filming can be done.

Stunt Coordinator: Individual who arranges and plans stunts.

Stunt Double: Stunt performer who specifically takes the part of another actor for a stunt.

Subcarrier: A secondary signal containing additional

information that is added to a main signal.

Subjective Shot: A camera shot which appears to take the viewpoint of an actor.

Submaster: A volume control which enables an operator to group a number of different audio or lighting sources together and then to control them with one fader.

Subpicture Menu: Menu used to select a subpicture stream.

Subpicture: Graphic bitmap overlays used in DVD-Video to create subtitles, captions, karaoke lyrics, menu highlighting effects, and so on.

Subplot: Also called the "B Story," the subplot is used in various ways, weaving in and out of the main action.

Subsampling: Also called Downsampling, it's the process of reducing spatial resolution by taking samples that cover larger areas than the original samples or of reducing temporal resolutions by taking samples that cover more time than the original samples. See also Chroma Subsampling.

Substrate: The clear polycarbonate disc onto which data layers are stamped or deposited.

Subtext: Implicit meaning in a play or film which lies beneath the language of the text.

Subtitle: A textual representation of the spoken audio in a video program. Subtitles are often used with foreign languages and do not serve the same purpose as captions for the hearing impaired. See also Subpicture.

Subtractive Color: The process wherein, as white light passes through a piece of film, yellow, magenta, and cyan colors are filtered out, leaving the many colors of the spectrum.

Success D'estime: A production that is well received by critics but that does not achieve popular or financial success.

Successive Contrast: The tendency of certain complementary colors or tones to exaggerate each other when viewed in temporal proximity. For example, the green color of grass will be exaggerated if a scene with large areas of intense red, magenta, or blue precedes it.

Suicide: A term for a distribution wiring connector with male plugs at both ends.

Sun Guns: Camera mounted lights that are portable and provide constant and shadow-less illumination. The problem with sun guns is that often a scene illuminated with one light is very unpleasant to look at. Also a scene with no shadows does not look very real or 3-D. The most effective use of a sun gun is to use it off camera to the left or right side.

Super 16: A format using single perf 16mm film on which a wider image is exposed than is the case with regular 16mm, using the area that would normally have the soundtrack. Super 16mm was conceived specifically for blow up to 35mm, and is typically rather inconvenient for anything else.

Super Speed: Another way to describe a fast prime lens, typically with a T-stop of 1.3.

Super VideoCD: Next generation VideoCD, defined by the China National Technical Committee of Standards on Recording, that hold 35-70 minutes of digital audio and video information. MPEG-2 video is used, with a resolution of 480 x 480 (29.97 Hz frame rate) or 480 x 576 (25 Hz frame rate). Audio uses MPEG-1 layer 2 at a bit rate of 32-384 Mbps, and supports four mono, two stereo, or 5.1 channels. Subtitles use overlays rather than subpictures (DVD-Video) or being encoded as video (VideoCD). Variable bit-rate encoding is used, with a maximum data bit rate of 2.6 Mbps. IEC 62107 defines the Super VideoCD standard.

Superblack: A keying signal that is embedded within the composite video signal as a level between black and sync. It is usually used to improve luma self-keying because the video signal contains black, making a good luma self-key hard to implement. When a downstream keyer detects the super black level, it inserts the second composite video signal.

Supercardioid Microphone: A moderately-directional microphone response pattern.

Superconductors: A general classification of experimental conductors that exhibit zero or near-zero resistance to the passage of an electrical current.

Superimposition: The simultaneous showing of two pictures in the same area of a video screen.

Supertelevsion Station: A term originally used to describe WTBS in Atlanta; supertelevision station is now used to describe any broadcast television station that has its signal distributed nationally via satellite or cable (i.e., WGN in Chicago, WSBK in Boston, WWOR and WPIX in New York, etc.).

Supplementary Lenses: An optical attachment used with a video camera lens that alters its optical characteristics. Three general categories are used: those that enable a camera lens to focus on objects at very close distances, those that increase the effective focal length of the lens (creating a telephoto effect), and those that decrease effective focal length (providing a wide-angle effect).

Supporting Feature: A feature film which appears (typically in a double-bill) with a feature presentation.

Surrealism: An avant-garde movement in the arts during the 1920s which endeavored to re-create unconscious experience with shocking, dreamlike images. Surrealistic films rejected traditional notions of causality and emphasized incongruous, irrational action instead.

Surround Sound: A sound system which creates the illusion of multi-directional sound through speaker placement and signal processing. See also Dolby, SDDS, DTS, and THX.

Sustaining Program: A television program not supported by advertising. Commonly, public service programming.

SVGA (Super VGA): An analog computer video signal format with 600 scan lines each normally representing 800 pixels across. We believe that this format was defined based on a video memory size of half a megabyte for 256 colors, or one 8 bit byte per pixel.

S-VHS (Super VHS): An improved version of the VHS tape format capable of recording better picture resolution (definition). A higher-density tape is required which provides a wider luminance bandwidth, resulting in sharper picture quality (> 400 horizontal lines vs. 240 for standard VHS) and improved signal-to-noise ratio. Because the equipment is usually smaller and lighter than 3/4" equipment, it is ideally suited for ENG/EFP applications.

S-Video (Separated Video): Describes a system of plugs and jacks used to interconnect camcorders, VCRs and TV monitors, which keeps the chrominance (color) and luminance (brightness) information separate. Also called Y/C connectors (luminance/chrominance), this system greatly improves picture quality by keeping any signal interaction (degradation) to a minimum.

SVM: See Velocity Scan Modulation.

Sweep Reversal: The reversal of electronic video scanning resulting in the left-to-right and/or top-to-bottom reversal of a television picture.

Sweep Set: Playing an uninterrupted set of music for a stated period of time is called a sweep set. Also called 10-in-a-row, continuous music, etc.

Sweep: A period of time (usually four weeks) during which all local TV markets are simultaneously measured and reported by a ratings service is called a sweep. Traditionally the sweep months are November, February, May, and July.

Sweetening: Electronically improving the quality of an audio or video signal, such as adding sound effects, audio ambience, or laugh tracks.

Swing Gang: Group within the art department that construct and take down a set.

Swish Pan: Panning the camera so rapidly across a scene that the image blurs on the screen. Also called flash pan, whip pan, zip pan.

Switchback: A camera shot that returns to the central action after a cutaway.

Switcher: General term for a device used to select different signals (audio, video, or RF) from various sources. It's also the technician who operates the television input control console to select or mix video input is called a switcher. They are also known as the Technical Director (TD).

SXGA: A video graphics resolution of 1280x1024 pixels.

Symbol: An object or image that has significance within a dramatic context beyond its literal meaning.

Symbolic Set: A nonrealistic setting intended to suggest an environment or setting—often somewhat abstractly.

Sync Beep (Sync Tone): In double system shooting with certain cameras, a tone feed into a magnetic tape recorder at the same time that a light in the camera exposes a few frames of film. The fogged section is later aligned with the beep tone to achieve synchronization of the sound to the picture.

Sync Buzz: An undesirable noise containing harmonics of 59.94 Hz. Heard on television audio under certain signal and transmission conditions as, for example, when the transmission of electronically-generated characters are of high level or have a resolution greater than the NTSC broadcast capability.

Sync Frame: Physical record unit of 1488 channel bits length comprising data (91 bytes) and a SYNC code. One physical sector consists of 26 sync frames.

Sync Generator: An electronic device that generates the variety of timing pulses needed to keep the video equipment synchronized.

Sync Mark: The point at which the clapsticks come together at the beginning of a shot, and the accompanying sound on the soundtrack. An "X" mark on a single frame at the beginning of a reel of picture that lined up with a second sync mark on a roll of sound (May also be used anywhere where needed). Sync marks are also used at the beginning of A&B rolls.

Sync Noise Gate: A sync noise gate is used to define an area within the video waveform where the video decoder is to look for the sync pulse. Anything outside of this defined window will be rejected. The main purpose of the sync noise gate is to make sure that the output of the video decoder is nice, clean, and correct.

Sync Pop: A single frame tone placed on the soundtrack so as to correspond with the "2" frame on the SMPTE leader.

Sync Restoration: A process which replaces distorted and missing sync information by checking incoming sync, analyzing the frequencies involved and generating new fully restored sync.

Sync Roll: The momentary or continuous vertical rolling of a television picture due to loss of sync.

Sync Sound: Sync sound is sound recorded while shooting picture. Usually it involves footage of people speaking, and is thus sometimes called lip sync. It must be recorded with either crystal or cable sync to line up and not drift out of sync.

Sync Stripper: A video signal contains video information, which is the picture to be displayed, and timing (sync) information that tells the receiver where to put this video information on the display. A sync stripper pulls out the sync information from the video signal and throws the rest away.

Synching Dailies: The picture and sound workprints of a day's shooting assembled for synchronous interlock.

Synchronization (Sync): A term used in electronics to

describe the precise alignment of two signals or functions. In video, sync is an essential element for maintaining the proper clocking of video signals. The sync signal is used by a monitor to know where and when to draw the on-screen video image. Sometimes referred to simply as "sync."

Synchronization License: A step in obtaining clearance to use a copyrighted work which grants a producer permission to use the music for a specific period.

Synchronizer: A very helpful tool of the editing room, a synchronizer is a device with a center axle and several sprocketed wheels attached to it. The wheels are called gangs. Film may be clamped into the gang, so that it can be measured with a footage counter on the front of the synchronizer. One revolution of the synchronizer equals one foot of film. Several elements, such as film and sound, A&B rolls, can be run in tandem can easily cut to the same length. It is used by the negative cutter for the assembly of A&B rolls, as well as for logging, measuring footage, syncing, and checking sync in the editing room.

Syndex: See Syndication Exclusivity.

Syndicated Program: A program that is produced for national distribution, but which is shown on individual local television stations rather than on a national network is called a syndicated program. These programs may be sponsored either locally or nationally.

Syndication Exclusivity (Syndex): Syndication exclusivity provides exclusive rights for a local television station to air a syndicated program in its home market (DMA). It is used primarily as a safeguard against syndicated programs airing on supertelevision stations.

Syndication Financial Interest (Synfin): Synfin is a FCC rule that prohibits the broadcast networks from having any financial interest in the syndication of programs that have aired on their networks. The rule was repealed in 1996, allowing the networks to have a financial interest in syndicating their programs.

Syndicators: Companies that sell programs to individual television stations rather than to the networks are known as syndicators. Often the production company will have its own syndication company to sell television stations.

Synecdoche: Use of a part to represent the whole.

Synfin: See Syndication Financial Interest.

Synopsis: A summary of the major plot points and characters of a script, generally in a page or two.

Syntagmatic Structure: In semiotics, the manner in which signs are linearly and/or temporally organized. E.g., the batting line-up in baseball is in syntagmatic order. See also Paradigmatic Structure.

Syntax: The rules governing construction or formation of an orderly system of information. For example, the syntax of the MPEG video encoding specification defines how data and associated instructions are used by a decoder to create video pictures.

Synthesized Music: Music created entirely by electronic means without the use of traditional musical instruments.

Synthesized Stereo: An approach to creating a simulated stereo effect from one or more monophonic audio signals.

Synthesizer: An electronic device with associated software used to create sound effects.

System Menu: The main menu of a DVD-Video disc, from which titles are selected. Also called the title selection menu or disc menu. T Tera. An SI prefix for denominations of one trillion (10^{12}).

T

T-1 Circuit: A T-1 circuit or leased line equals 1,544,000 bits per second or 24, 56 kbs leased lines.

T-3: A T-3 Circuit (45,000,000 bits per second) is the minimum backbone speed of all major national ISPs in the United States.

Tachometer: A gauge on a camera measuring the film speed when the camera is running.

Tag: The dealer identification, usually added to the end of a commercial announcement that advises viewers or listeners where the product being advertised can be purchased in their market is called a tag.

Tail Leader: Leader used at the finish end of a strip of film.

Tail Slate: Sometimes it is necessary to mark a shot at the end rather than at the beginning. When this is done it is called a tail slate. It is customary to call "Tail Slate!" just before clapping the slate, so that the person syncing the film does not get confused. To easily distinguish a tail slate, the slate is held upside down when marking the shot.

Tail: The end of a shot or a roll. The inclusion of tails creates a dirty EDL. When used in a true random access edit suite the tails give room for edits to be adjusted without the need to re-record footage.

Take Down: Reducing the light on an object by means of nets, scrims, dimmers or wasting light.

Take Up Reel: An empty reel, used on a projector to gather up the film after it has passed through the movement.

Take Up Spool: An empty spool in a camera used to gather up the film after it has passed through the movement.

Take: The shot resulting from one continuous run of the camera. A filmmaker generally films several "takes" of the same scene and then selects the best one.

Talent Release: A form signed by a person appearing in a videotape which grants legal permission to a production agency to broadcast the segment under specified terms.

Talent: Generic term for the people or creatures assuming primary on-screen roles in a videotaping.

Talking Head: Slang for the typical (and not too exciting) head-and-shoulders shot of actors seen on talk shows and newscasts.

Tally Light: Automatic indicators on camera front and within viewfinder that signal recording in progress: seen by both camera subject(s) and operator.

TAP (Total-Audience-Plan) Plan: Usually a TAP Plan means that spots are evenly divided among premium time periods (i.e., in radio, a TAP Plan might be three spots in morning drive, three in midday and three in afternoon drive). A television station offers an attractive rate to an advertiser who purchases a scheduled on a TAP Plan.

Tape Format: Any of several tape widths and recording methods, such as Digital Betacam, S-VHS, or D-5, used in video recording systems.

Tapeless Work Station: An audio or video editing console or work station which typically uses computer disks as a recording medium.

Target Audience: The intended audience for a production.

Target: The light-sensitive front surface of a television camera imaging device. It coincides with the focal plane of the lens.

TBC: See Time Base Corrector.

TCP/IP: "Transmission control protocol/Internet protocol." TCP/IP is a combined set of protocols that perform the transfers of data between two computers.

TD: See Technical Director.

TDF: Telediffusion de France.

TDL (Telecine Decision List): A list of the edits made in a telecine session which can be loaded into an off-line editor.

TDM: See Time Division Multiplex.

Tearing: A video scanning aberration where the edges of an object appear rough and jagged. Often associated with chroma key problems.

Tease: A short, initial, often provocative segment of a production intended to grab and hold attention. On television news, brief announcements of upcoming stories used to maintain viewer attention.

Technical Advisor: Individual with expertise in a particular field who provides advice

Technical Continuity Problem: Any one of several types of unintentional changes in video or audio quality. Examples are unintended changes in color balance or changes in audio levels during the course of consecutive shots.

Technical Director (TD): The individual who operates the control room switcher and is in charge of various technical aspects of a production.

Technical Working Group (TWG): A general term for an industry working group. Specifically, the predecessor to the CPTWG.

Technicolor: A type of color film process, used mostly from the late 1930s to the 1950s.

Telecine Artist: The operator of a telecine machine. Also called a Colorist.

Telecine Converter: Imaging device used in conjunction with a movie projector and camcorder to transfer film images to videotape.

Telecine: An area of a production facility with equipment for converting film and slides to video.

Teleconferencing: Closed-circuit television productions which link two or more locations and are designed to facilitate discussions and interaction between professionals.

Telephoto Lens: Any lens with a focal length significantly greater than the standard focal length for the format in which it is to be used. The standard length for 35mm still cameras and one inch television cameras is 50mm.Used as an equivalent to Long Lens, but for those who wish to be overly exact, a telephoto lens is a long lens that is physically shorter than its focal length.

Teleplay: Script written to be produced for television whose length is 42-48 long for a standard one hour TV show, often structured in the teaser/4-acts/tag format. A 1/2 hour sitcom has an average page count of 45-50 pp double spaced and usually consists of 2 acts with 4-6 scenes per act.

TelePrompTer: Originally, a brand name for a camera prompter. A device used by on-air talent that rolls an image of a script across a screen near the camera lens.

Telescriptions: Produced by Louis Snader in the 1950s, short films of musical performances that were marketed to television stations for use in variety shows or as filler material.

Teletext: A method of transmitting data with a video signal. ITU-R BT.653 lists the major teletext systems used around the world, while ETS 300706 defines in detail the teletext standard for PAL. North American Broadcast Teletext Specification (NABTS) is 525 line system C. For digital transmissions such as HDTV and SDTV, the teletext characters are multiplexed as a separate stream along with the video and audio data. It is common practice to actually embed this stream in the MPEG video bitstream itself, rather than at the transport layer. Unfortunately there is no wide-spread standard for this teletext stream—each system (DSS, DVB, ATSC, DVD) has its own solution.

Television Advertising Bureau (TVB): The TVB is a corporation sponsored by various networks and television stations, whose purpose is to promote the use of televisions as an advertising medium.

Television Apparatus: The combined work of all of the various factions (bankers, media corporations, directors, scriptwriters) that create television programs and the viewing experience itself—including the psychological mechanisms at work during TV viewing.

Television Criticism: Non-empirical, analytical methods (e.g., auteurism, genre study, semiotics, and feminism) employed to understand systems of meaning on television. The term is also used in the popular press to refer to evaluative reviewing of television.

Television Movie: A television production of a singular event (such as an awards show or concert) as opposed to a regularly scheduled series.

Television Production: The process of creating television programming.

Television Station Break: The time allotted between network and local programs where commercial announcements of varying lengths are scheduled and where the television station identifies its call letters and location.

Television Station Identification: The announcement between programs on the hour or half-hour in which local television stations are required by the FCC to identify themselves by call letters and city of origin is known as television station identification.

Televisual: Characteristic of television.

Telewriter: A writer who either adapts an existing work for production on television, or creates a new teleplay.

Temp Dub: A preliminary mixing of dialogue, music, and sound effects, usually so that a first cut may be viewed with all of these elements incorporated.

Temp Programs: Temporary jobs that may be used for special applications are known as temp programs.

Tempo Mapping: Using a computer screen representation of a musical segment, complete with discernible music measures, to create electronic music and effects for a production.

Temporal Aliasing: A visual defect that occurs when the image being sampled moves too fast for the sampling rate. A common example is wagon wheels that appear to rotate backwards.

Temporal Compression: A compression technique that operates across time by eliminating similar data from frame to frame. For example, the first of two frames is spatially digitized in its entirety, and in the next frame, only the information that has changed is digitized. Temporal compression reduces the amount of data needed to describe a given frame.

Temporal Resolution: The clarity of a moving image or moving object, or the measurement of the rate of information change in motion video. See also resolution.

Temporal: Relating to time. The temporal component of motion video is broken into individual still pictures. Because motion video can contain images (such as backgrounds) that do not change much over time, typical video has large amounts of temporal redundancy.

Temporally Adjacent: Refers to scan lines transmitted consecutively. The term is needed in discussing interlaced video where two temporally adjacent lines are both odd or both even. For progressive scanned video, lines that are temporally adjacent are also spatially adjacent.

Tenner: A standard studio 10K lighting unit, as opposed to a baby 10 or a Big Eye, which are also 10K lighting units.

Terabyte: 1 trillion bytes.

Termination Switch: A switch that connects and disconnects a load resistance to a video input, used to terminate the line. In order for a video signal to be correctly transmitted without loss, proper end of line impedance is essential. A 50 or 75 ohm resistor is usually employed to accomplish this. When the termination switch is off, the unterminated video signal is looped to the next device where the signal can be transmitted in parallel. The final device in the chain must be terminated using the termination switch.

Termination: The resistance or load that marks the end of coaxial video conductors usually 75 ohms. A lack of termination generally causes video level problems.

Terrestrial Interference: Electrical noise from earth-based RF that interferes with satellite reception.

Terrestrial: A broadcast signal transmitted "over the air" to an antenna.

Tessellated Sync: This is what the Europeans call serrated sync. See also Serration Pulses and Sync.

Test Pattern: Any number of standardized electronic patterns or camera charts intended to evaluate specific video qualities such as linearity and resolution.

Test Tone: A zero-decibel tone used as a reference standard.

Text: A segment of the televisual flow, such as an individual program, a commercial, a newscast, even an entire evening's viewing.

TFN (Until Further Notice): The abbreviation for Until Further Notice, this refers to orders having a start date but no specified end date (TFN orders).

The Call: This is the sequence of directions that begin a take, typically: "Roll Sound!" "Roll Camera!" "Mark it!" and Action!" (21)

The Groundglass: A flat surface of etched glass in the viewfinding system of a camera that is the same distance from the lens as the film plane.

The Movement: The parts of a camera or projector that move the film intermittently: the pull-down claw, the rollers before and after the loops, and the gears connecting these parts form the movement. If there is a registration pin, this is also part of the movement. Sometimes the shutter can also be considered part of the movement.

The Slate: A board with two hinged sticks attached. The slate is used to record a scene number and sync point (via the clapstick) at the beginning of a shot.

The Taking Lens: On a turret, the lens that is actually in front of the gate, producing an image on the film. This is also a term used to refer to on-camera subjects in a video production.

Theatrical film: Films originally designed to be shown in theaters, as opposed to made-for-TV films (MOWs).

Thematic Editing: See Montage Editing.

Thirty-degree Rule: A principle of visual continuity which states that the relative angle of view of any two similarly framed shots of the same subject which are to be cut together should vary by at least thirty degrees.

Three Shot: As with the two shot, the conventional framing of three characters in a medium shot.

Three-act Structure: Traditional storytelling sequence which includes the set-up, the complication, and the resolution.

Three-camera Format: Standard video-studio setup which utilizes three different cameras.

Three-chip Camera: A video camera that uses three CCDs, one devoted to each of the primary colors.

Three-dimensional (3-D) Animation: Building a 3-D computer model of an object on a video screen and thereafter being able to rotate the object in several directions.

Threefold Set: A free-standing, three-section set, hinged in three places.

Three-point Lighting: Basic lighting approach employing key, back, and fill lights to illuminate subject with sense of depth and texture. Strategic placement imitates natural outdoor lighting environment, avoids flat lighting.

Three-two Pull-down: A technique for converting the 24-picture-per-second motion picture standard into a 30-frame-per-second TV system. Although more than one technique is used, typically, one film is scanned three times by the TV system and following picture is scanned twice.

Threshold: The measure of sensitivity of a satellite receiver measured in decibels (dB).

Thumbnail: A low-resolution version of an image that gives the viewer a quick reference to what the image will look like full size.

THX: A theatrical film exhibition sound system which maintains a consistent sound standard from theatre to theatre. Generally, this system uses Dolby Stereo Surround as the basis of the exhibition standard, although a number of other systems have also now qualified.

Ticker: Information moving across the bottom of the screen—such as sports scores and weather updates.

Tie In: A power feed obtained by temporarily clipping on to the main service of a location. This methodology is illegal in many areas.

Tiering: Subscribers to a cable system with one or more programming services beyond the basic offerings pay an extra charge. Hence, each additional price increment (or service) is referred to as a tier.

TIF (Tag Image File Format): A common graphic file format supported by most DVD-Video authoring tools for the importation of still images, menu backgrounds, and highlight overlay images.

Tight Wind Hub: A tight wind is useless without it. This is the hub that holds a core on the spindle of a rewind.

Tight Wind: A handy attachment sometimes found on an editing bench on the right rewind, used to wind film onto a core and giving it a very smooth edge. It can be quicker than opening and tightening split reels if you are just rewinding an entire roll.

Tighten Up: To zoom in or dolly in on a subject.

Tilt: A shot where the camera pivots vertically, turning from top to bottom or from bottom to top.

Timber: Sound characteristics, including overtones, which differentiate musical instruments.

Time Base Corrector (TBC): An electronic device that stabilizes the scanning and timing pulses in video. Commonly used to bring technical attributes of the video from small field cameras and recorders up to broadcast quality.

Time Base Signal: A signal recorded on the edge of film in a camera to match a signal recorded on a magnetic recording which is used as a fast means of synchronizing film and sound workprints.

Time Code Lock: To synchronize the built-in time code generator of video equipment such as a VTR to an external time code.

Time Code: Information recorded with audio or video to indicate a position in time. Usually consists of values for hours, minutes, seconds, and frames. Also called SMPTE time code. Some DVD-Video material includes information to allow the player to search to a specific time code position.

Time Compressor: Electronic signal processing system allowing a recorded videotape to be replayed somewhat faster or slower than normal without noticeable effects.

Time Division Multiplex (TDM): The management of multiple signals on one channel by alternately sending portions of each signal and assigning each portion to particular blocks of time.

Time Lapse Photography: Form of animation in which numerous single frames are filmed spaced at a given interval to show a process that would take a very long time to occur.

Time Lapse Recording: Periodically videotaping a minimal number of frames over long durations of actual time. Upon playback, slow processes such as a flower blooming may be viewed in rapid motion.

Time Lapse: Time lapse is when single frame shooting is used to dramatically speed up the action over the course of a long period of time. Typically it is a process where a single frame is shot after a consistent pause. It could be one frame every ten seconds, or one frame every hour, and such.

Time Line Editing: A computer-based method of editing in which video and audio clips are represented on a computer screen by bars proportional to the length of the clip. These bars can be moved and resized along a grid whose horizontal axis relates to the time of the program.

Time Period: An interval of time usually defined by day(s) and hour(s) without regard to the programming aired is known as a time period.

Time-Based Correction: The process of correcting picture jitter, skewing (shifting of the video image), or color breakup caused by reproducing the video signal from tape.

Time-code Generator: A device which supplies an electronic SMPTE/EBU time-code signal to recording equipment.

Timed Print: Unlike a One Light Print, this is a print where the timer has gone through and timed every shot.

Timeline: In nonlinear editing, the area in which audio and video clips are applied, typically giving duration in frames and seconds.

Timeliness: A measures of newsworthiness relating to how recent or new the developments in a story are.

Time-motion Study: Similar to time-lapse photography except that the condensation of time is designed to allow researchers to analyze (generally human) actions over a period of time in an effort to minimize wasted or inefficient movement.

Timer: The person at the lab who goes through the film, shot by shot and selects the printing lights.

Timing Lights: These are the lamps of the contact printer at the lab. Their brightness can be controlled, which is measured in a scale of 1 to 50, 1 being the darkest and 50 the brightest. In color there are three lights used together: Red, Green, and Blue. When working with negative it is worth remembering that the values are reversed: the brighter the light, the darker the print will be. Also called Printing Lights.

Timing Report: A list of the timing lights and corresponding footages the lab used in making the print. The timing report can be very helpful for analyzing the footage and judging the possibilities of further corrections. Any serious problems with the footage (out of focus, scratches, edge fog, etc.) are usually also noted on the timing report.

Timing: The lab's process of selecting printing lights to for the proper rendition of exposure and color when making a print. The term is a little consuming, as it has nothing at all to do with "time" as in "running time" or such.

Tinting: Immersing positive film stock in a dye which is absorbed by the gelatin of the emulsion, producing a uniform single color.

Titillation Component: A measures of newsworthiness relating to the sexual or sensual components of a story.

Title Design: The manner in which title of a movie is displayed on screen is widely considered an art form.

Title Key: A value used to encrypt and decrypt (scramble) user data on DVD-Video discs.

Titles: Graphic information appearing at the beginning of a program, generally including the title, author, producer, writer, director, and major personalities. Titles may be added to a video scene during shooting or in post-production. Sophisticated titling devices allow the user to prepare text and graphics in various sizes, fonts and colors to be triggered later, one-by-one, at appropriate places within a production. Many video cameras include basic titlers or permit externally-generated titles to be mixed with the video image during shooting.

Toe: In television, the bottom portion of a gray-scale gamma curve representing the reference black and the darker portions of the gray scale.

Tonal Compression: Commonly, a darkening of skin tones associated with either underexposure, or exceeding the optimum contrast ratio of a video system. The inability of a video system to differentiate between different gray scale reflectance values.

Tonal Merges: An undesirable visual blending of different objects in such a way that they cannot be clearly distinguished.

Tone Generator: Electronic circuit which can create an even, standardized reference tone for test and setup purposes.

Tone: A 1,000 Hz sine wave used at the beginning of a tape to provide consistent volume when transferring sound.

Tongue: To move a camera boom to the left or right.

Toning: Chemically treating the silver salts in undeveloped, negative film stock that, once developed, changes the black portions of the image into another color. The difference between tinting and toning is that tinting produces a uniform overall color while toning only changes the black silver image into another color.

Top Lighting: Light source above the subject, leaving shadows below.

Top of the Hour: The top of the hour refers to the beginning of an hour, as on a clock face (e.g., 12: 00, 3: 00, etc.).

Total-Audience-Plan: See TAP.

Trace Capability: The ability of some editing systems to retain and display the original time-code references on footage after the codes have been changed during later stages of editing. This capacity can be important in going back to original footage and locating needed segments.

Track Buffer: Circuitry (including memory) in a DVD player that provides a variable stream of data (up to 10.08 Mbps) to the system decoders of data coming from the disc at a constant rate of 11.08 Mbps (except for breaks when a different part of the disc is accessed).

Track Pitch: The distance (in the radial direction) between the centers of two adjacent tracks on a disc. DVD-ROM standard track pitch is 0.74 mm.

Track Sampling: The process of selecting segments of audio piece for demo purposes.

Track: A single component or channel of a soundtrack. See also Sound Mix.

Tracking Shot: A tracking shot is one where the camera is placed on a dolly and is moved while filming. Also known as a dolly shot.

Tracking: An adjustment to the playback heads of a videotape machine to make them match the phase of recording heads. Sometimes the adjustment is necessary because the tape was recorded on another machine. Tracking problems cause video breakup and aberrations at the top or bottom of the picture.

Trade Libel: A false or grossly misleading statement which damages a business or product.

Trade: An agreement between a television station and an advertiser, by which the television station receives merchandise or services, not cash, in exchange for airtime. For example, a retailer might not be charged for a spot if they gave prizes to the television station for a promotion. Outside the U.S. trade is commonly referred to as "contra."

Trades: Newspapers that report the daily or weekly

entertainment news of the entertainment industry; The Hollywood Reporter, Daily Variety, and Weekly Variety.

Traffic Department: The department in a television station or a network that coordinates the information flow between programming, sales, and continuity, and uses that information to create the program log is known as the traffic department.

Trailer: An advertisement for a movie which contains scenes from the film. Historically, these advertisements were attached to the end of a newsreel or supporting-feature, hence the name. Doing this reduced the number of reel changes that a projectionist would have to make.

Trailing Dots: Cross-luminance near sharp saturated moving color edges due to the failure of the chrominance filter in the temporal dimension and thus providing incomplete Y/C separation.

Trance Film: A type of expressionistic film in which a character, frequently the filmmaker, experiences a personal revelation by playing out a psychological drama in dream-like surroundings.

Transcoding: The process of converting from one production format to another. This may include up conversion, down conversion, translation, de-interlacing, component to/from composite coding and time compression/expansion.

Transfer Rate: The speed at which a certain volume of data is transferred from a device such as a DVD-ROM drive to a host such as a personal computer. Usually measured in bits per second or bytes per second. Sometimes confusingly used to refer to data rate, which is independent of the actual transfer system.

Transform: The process or result of replacing a set of values with another set of values. A mapping of one information space onto another.

Transition: See Bumper.

Translation: A process that converts an image stream from one temporal sample structure to a higher or lower temporal structure. i.e. 3**:** 2 pull down to convert film (24fps) to video (30fps).

Transmission Standard: The broadcast standard used for transmitting signals to the home.

Transmitter: In television, an electronic device that modulates the audio and video signals onto a carrier wave for broadcast.

Transponder: Equipment inside a satellite, responsible for receiving a single uplinked channel and re-broadcasting it back to earth.

Transport Stream (MPEG): A bit stream packaging elementary streams together. It can accommodate several programs with independent clocks. Transport streams have fixed length packets and are intended for transmission over error prone channels. This stream is often called simply the MPEG signal.

Transportation Coordinator: Individual responsible for managing drivers and coordinating the transportation of a production's cast, crew, and equipment from the various locations and sets used for filming.

Traveling Matte: A process shot in which foreground action is superimposed on a separately photographed background by an optical printer.

Traveling Shot: A shot in which the camera, mounted on a vehicle, moves while filming. Traveling shots are sometimes identified more specifically according to the kind of vehicle used to move the camera (a dolly shot or a trucking shot, for example). When tracks are laid down for the camera to roll on, the shot is usually called a tracking shot.

Treatment: A written description of a film story which may later be developed into a script. Also called a Film Treatment.

Treble: Sounds representing the higher audio frequencies, generally from about 4,000 Hz up.

Triad: Three-color phosphor dots or stripes on the faceplate of a tri-color CRT.

Triax: A coax-type video cable with three conductors.

Trick Shot: A film or a shot created by special camera or optical techniques, such as stop-motion photography or double exposure. See also Pixilation.

Tri-level Sync: A sync signal that has three levels, and is commonly used for analog HDTV signals.

Trilogy: A series of three movies that are closely connected by plot. Often, a storyline from the first film of a trilogy is altered, twisted or modified by the second or third part of the series.

Trim Bin: A bin on wheels lined with a fabric bag and topped off with a frame with a row pins on which to hang film while editing. Oddly enough, a trim bin is not used for trims, which are small, but for selects and outtakes.

Trim Control: A control on an editor which makes it possible to add or delete a specified number of frames to an edit point.

Trim: Changing an in- or out-point of a shot. In linear systems this becomes difficult if further edits have been made as it can change the duration of the program. Nonlinear systems typically offer comprehensive trimming options to either the outgoing or incoming video and/or audio, separately or together. The flexibility of true random access suites means trims are never a technical problem and can always be completed quickly, no matter how far back they are in the program. See also Crop.

Trimmer: A kind of variable capacitor, often used to fine tune pushbutton TV and radio tuners, or sometimes used as the screwdriver adjustments behind a TV to adjust convergence, pincushion distortion, etc.

Triple Re-entry: A video switcher system in which the results of one mix-effects bank can be directed through another, and then the results of that can, in turn, be directed through a third mix-effects bank, thus combining three layers of video effects.

Tripod Head: The part of the tripod with the pan and tilt mechanism to which the camera is attached.

Tripod: Three-legged camera mount offering stability and camera placement/movement consistency. Most are lightweight and used for remote recording.

Tri-standard: A TV or VCR which can work with an NTSC, PAL, or SECAM TV signal.

Tristimulus: A three-valued signal that can match nearly all colors of visible light in human vision. This is possible because of the three types of photoreceptors in the eye. RGB, YCbCr, and similar signals are tristimulus, and can be interchanged by using mathematical transformations (subject to possible loss of information).

Trixel: The time taken to transfer a single pixel through the entire proposed data path. Assume to be large enough to measure on know time-scales.

Trombone: A tubular, extending device which is generally used for suspending lights from set walls.

Truck: A left or right movement of the camera along with its mount.

Trucking Shot: A camera move which films the subject from side to side. See also Traveling Shot.

True Color: True color means that each sample of an image is individually represented using three color components, such as RGB or Y'CbCr.

Truth: A legal defense against libel or slander. If the truth of a statement can be proven, it cannot be considered slanderous or libelous.

T-stop: Similar to an F-stop, some lenses, particularly zoom lenses, will have F-stops on one side of the aperture ring and t-stops on the other. To differentiate the two, the t-stops will be red and the F-stops white. T-stops are used in place of F-stops for setting exposure. Lenses with a lot of glass elements will often lose a little bit of light. The t-stops are calibrated to the actual amount of light that is hitting the film, rather than arrived at mathematically, as is the case with F-stops. However, the F-stops are still relevant, because while the t-stop should be used to set the exposure, the resulting F-stop will indicate how much Depth of Field you have.

Tuner: Television and VCR component that receives RF signals from an antenna or other RF sources and decodes into separate audio and video signals.

Tungsten: The color temperature of artificial light which is 3,200K on the color temperature scale. Quartz Lights use a tungsten filament, which burns at 3,200K, and gives us this term. Color film for indoor shooting is balanced for tungsten light, otherwise the image would appear orange in hue. If tungsten balanced film is used out-of-doors without a correction filter the image will have a washed-out blue hue.

Tuning Slug: A movable ferrite core used to adjust a coil's behavior (inductance). May be the screwdriver (sometimes hex wrench) adjustments in back of a TV to adjust convergence, etc.

Turnaround: A state of limbo that a movie enters after a studio decides to drop it. In turnaround, the producers have a chance to set the project up with another studio or with different talent. (210

Turnkey: A system or installation that is complete and ready to run without need of further additions or modifications.

Turret: A rotating lens mount allowing for the mounting of three or sometimes four lenses on a camera, allowing for the quick change from one lens to another. Only one is in use at any given time, and this is known as the taking lens.

Turtle: A flat, on the floor mount, for large lights with a junior receiver.

TV Black: The blackest part of a TV picture generally 3 percent reflectance.

TV Households: A statistical estimation of homes that have a working television is known as TV Households.

TV Q: A Qualitative Ratings study that measures the popularity of television personalities and television programs is known as a TV Q. It is also known as a "Q Rating."

TV Safe: The area of a filmed image which will normally appear on a home television set after a film has been transferred in a telecine and then transmitted.

TV White: The whitest part of a TV picture generally 60 percent or more reflectance.

TVB: See Television Advertising Bureau.

TVCR: A TV set with built in VCR.

TVL (Television Line): On a test pattern consisting of closely spaced parallel lines, a TV line is either one black line or one white space that separates two adjacent black lines. The term TV Lines was introduced as an advertising gimmick to make TV, which has significantly less resolution than photographic film, seem to have more resolution than it does. Photographers, when they refer to lines of resolution on film, count only the black lines. Resolution expressed as TV lines (or as EIA) refers to the maximum number of them that can be side by side and still distinguishable, in the largest circle that in turn fits in the screen.

Tweak: To accurately align electronic equipment.

Tweco: A device similar to a slip pin connector, which is used on a feeder cable, and which has a positive twist connection.

Tweening: A process in animation by which frames are created that constitute a character's movement. These frames come in between the key frames the animator has designed and can be automatically generated by a computer.

Tweeter: A small speaker designed to reproduce high audio frequencies.

TWG: See Technical Working Group.

Twin Lead: See 300 Ohm above.

Twisted Pair: A standard telephone wire.

Two Shot: The framing of two characters in a medium shot.

Twofold Set: A free-standing, two-section set, hinged at the mid-point.

Twos Complement: The number calculated so that each bit of a binary number is inverted (ones are replaced with zeros and vice versa), then one (=000.0001b) is added ignoring the overflow.

Type-B Videotape Machine: An approach to 1-inch videotape recording used in Europe. It uses multiple rotating heads and records each complete video picture in segments.

Type-C Videotape Machine: An approach to 1-inch videotape recording used in NTSC counties. A complete television picture is recorded with each helical scan of the tape heads.

Typecasting: Selecting an actor or actress for a film role because of his or her physical type, manner, or personality, or according to a public image created by previous roles he or she has performed.

U

U, V: The video signal components B-Y and R-Y, respectively after they are modulated onto a color subcarrier suitable for subsequent construction of a composite video signal. To produce the picture, the luminance signal Y is also needed. U and V are used a lot for PAL composite video. Use of U and V in NTSC composite video is done but it does not quite optimize the color horizontal resolution to the maximum sensitivity of the human eye to reddish orange and greenish blue.

UDF (Universal Disc Format): A standard developed by the Optical Storage Technology Association designed to create a practical and usable subset of the ISO/IEC 13346 recordable, random-access file system and volume structure format.

UDF Bridge: A combination of UDF and ISO 9660 file system formats that provides backward-compatibility with ISO 9660 readers while allowing full use of the UDF standard.

UHF (Ultra High Frequency): Comment: Channels 70-83 are no longer used for TV broadcast and TV sets are no longer required to receive them. Other services such as cellular telephones use this part of the frequency spectrum, which as a whole extends from about 316 MHz to 3.16 GHz.

UHF Remote: Ultra High Frequency remote control that can operate the receiver from another room. The IR (Infra Red) remote needs to be pointed at the receiver.

Ultra-directional Microphone: A microphone with a highly directional response pattern.

Ultra-Sonic Cleaner: A sophisticated cleaning machine found at labs to clean negatives prior to printing or transfer to video. It uses sound waves to shake loose dust.

Ultra-Sonic Splicer: An expensive and sophisticated splicing machine used for splicing Polyester Base stock.

U-Matic: A VCR format that uses 3/4 inch wide tape and is somewhat similar to the now obsolete Beta format.

Umbrella: What the name implies, lighting accessory available in various sizes usually made of textured gold or silver fabric. Facilitates soft, shadow-less illumination by reflecting light onto a scene.

Unbalanced Line: Audio sources associated with non-professional equipment which rely on two-conductors. Unbalanced lines tend to be susceptible to hum and electronic interference. See also Balanced Lines.

Uncommitted Editing: Editing where the decisions are made and the edits completed but any can still be easily changed. This is best provided in a true random access edit suite where the edits need only comprise the original footage and the edit instruction. Nothing is re-recorded so nothing is committed. This way, decisions about any aspect of the edit can be changed at any point during the session, regardless of where the changes are required.

Under Lighting: Light source below the subject, leaving shadows above.

Undercrank: To run film stock through the camera slower than the standard speed of 24 frames per second (fps), producing fast motion on the screen when the film is projected at standard speed. See also Overcrank.

Underexposure: Filming a scene with less light than the emulsion of the film needs for a correct exposure. The image will be too dark. If compensated for in printing, the image will appear grainy, and very muddy.

Underground Film: An independent film which emphasizes the filmmaker's self-expression rather than commercial success. Underground films frequently challenge or experiment with traditional cinematic form and technique, hence they are also called avant-garde or experimental films.

Underscan: Video display scanning technique in which the entire video signal, including black parameter retrace portions, is displayed on the CRT.

Underscore: Music that provides emotional or atmospheric background to the primary dialog or narration onscreen.

Unidirectional Microphone: A highly selective microphone pickup pattern that rejects sound coming from behind while absorbing that from in front.

Unidirectional Prediction: A form of compression in which the codec uses information only from frames that have already been decompressed.

Union Scale: A basic salary for specific jobs set by a union.

Unipod Monopod: A camera support consisting of a single, adjustable leg.

Unique Selling Proposition (USP): Rosser Reeve's term for that certain something that distinguished one product from all the others.

Unit Production Manager: Executive who is responsible to a senior producer for the administration of a particular movie.

Unit: A slot within a break in which a commercial or non-commercial (PSA or Promo) announcement may be scheduled is known as a unit.

Unity Gain Amplifier: An amplifier or amplifier stage that doesn't actually amplify, namely the output is the same as the input. It does have useful purposes, for example to prevent signals internal to a system from leaking out of an input jack or port, or to convert from a low voltage high current circuit to a high voltage low current circuit or vice versa.

Unity: The number 1, when used as a multiplier.

Universal DVD Player: A DVD player that can play both DVD-Video and DVD-Audio discs.

Universal DVD: A DVD designed to play in DVD-Audio and DVD-Video players (by carrying a Dolby Digital audio track in the DVD-Video zone).

Universe Estimate: The total households/persons in a given demographic or population cell is known as the universe estimate. Network, barter syndication and local television markets have their universe estimates updated annually (in September). All cable networks have their universe estimates updated monthly.

Unsqueezed Print: A print in which the distorted image of an anamorphic negative has been corrected for normal projection.

Up Conversion: A process that resamples an image in order to produce an image with more pixels than are in the original image.

Up-cut: Switching to a videotape late, resulting in momentary loss of video or audio at the start.

Upfront Buying: The buying of national television advertising time for a full broadcast year (generally September through August) via one negotiation is known as upfront buying. Upfront buying generally requires representation throughout all four quarters. It allows cancellation options in the last six months of a buy and generally allows audience guarantees to advertisers.

Uplink: The carrier used by Earth television stations to transmit information to a satellite.

User Bits: Additional digital information that can be recorded within the SMPTE/EBU time code. A limited number of user bit letters or numbers can be entered to register reel number, date scene, take, etc.

User Data: The data recorded on a disc independent of formatting and error-correction overhead. Each DVD sector contains 2048 bytes of user data.

Uses-and-gratifications: A research method that sees the viewer as an active user, and attempts to chart the way that viewers employ television; this method quantifies how television fulfills viewers' emotional or intellectual needs.

USP: See Unique Selling Proposition.

UV Filter (Ultraviolet Filter): Transparent filter that absorbs UV wavelengths. Used to help penetrate haze and give clearer video of distant views. Also used over a lens simply to protect the surface of the lens.

UXGA: A video graphics resolution of 1600x1200.

V

V Chip: See EIA-744.

V: An IMDb notation to indicate that a particular title was originally released on video without a screening or being broadcast.

Vaporware: Software or hardware that is talked about, but may never actually appear.

Vari Speed: A motor or the control for a motor which will run a camera or an editing machine at speed faster or slower than sound speed.

Variable Bit Rate Encoding (VBRE): As opposed to constant bit rate (CBR), VBR allocates additional bandwidth to sections of encoded media that require more complex motion and higher-quality sound. VBR produces superior encoded output compared to CBR encodes with similar bit rates.

Variac: A simmer that reduces the voltage. It stands for VARIable AC.

Vaudeville: Anti-naturalistic performance style in which the actor reminds the viewer that the character is not a real person, often by directly addressing the viewer.

Vault Box: A white, flat, square cardboard box designed to hold 1,000 feet of 35mm or two 1,000 foot rolls of 16mm.

VBI (Vertical Blanking Interval): The scan lines in a television signal that do not contain picture information. These lines are present to allow the electron scanning beam to return to the top and are used to contain auxiliary information such as closed captions.

VBR (Variable Bit Rate): Data that can be read and processed at a volume that varies over time. A data compression technique that produces a data stream between a fixed minimum and maximum rate. A constant level of compression is generally maintained, with the required bandwidth increasing or decreasing depending on the complexity (the amount of spatial and temporal energy) of the data being encoded. In other words, data rate is held constant while quality is allowed to vary.

VBRE: See Variable Bit Rate Encoding.

VBScript: A proprietary Visual Basic-based programming language defined by Microsoft for use in their Internet Explorer Web browser.

VBV (Video Buffering Verifier): A hypothetical decoder that is conceptually connected to the output of an MPEG video encoder. Provides a constraint on the variability of the data rate that an encoder can produce.

VBV (Video-Black-Video): A preview mode used in conventional editing, that displays recorded video, a black segment, then recorded video again. This is useful for checking the exact frames of the original that will be visible after an insert edit. This is less relevant in nonlinear editing systems which provide comprehensive edit point viewing and trimming facilities.

VCAP (Video Capable Audio Player): An audio player which can read the limited subset of video features defined for the DVD-Audio format. (Contrast with universal DVD player.)

VCR: A common household appliance for recording and/or playing prerecorded videotapes.

Vector Graphics System: A graphic 3-D modeling electronic graphics approach in which the appearance of three-dimensions is created by being able to rotate or move the subject matter. The moving perspectives are automatically calculated and created by the software involved.

Vectorscope: A special type of oscilloscope designed to display the saturation and hue of chroma signals in a polar pattern. This device is essential in the evaluation of color signals when aligning color picture sources or matching and timing color sources in a television system.

Veeder Counter: A trade name for a mechanical counter which is actuated using a rotary shaft movement which can be calibrated to any system of measurement such as feet or frames.

Velocity Microphone: See Ribbon Microphone.

Velocity Scan Modulation: Commonly used in TVs to increase the apparent sharpness of a picture. At horizontal dark-to-light transitions, the beam scanning speed is momentarily increased approaching the transition, making the display relatively darker just before the transition. Upon passing into the lighter area, the beam speed is momentarily decreased, making the display relatively brighter just after the transition. The reverse occurs in passing from light to dark.

Verisimilitude: The impression of truth or reality.

Versioning: The practice, especially in news, commercials and promotions work, of making multiple versions of an edit for use in different circumstances. Nonlinear systems allow the editor to make different versions very quickly compared to linear editing. This can extend to complex, multi-layered edits which can have new stills or clips inserted. Thus there is no pressure to limit the original look to make new versions simpler.

Vertical Blanking Interval: See VBI.

Vertical Fader: See Linear Fader.

Vertical Filtering: Video production method used to reduce flickering of very thin horizontal lines or objects particularly when they are moving up or down. It consists of blending the content of adjacent scan lines which of course reduces vertical resolution. It can be done optically, for example the telecine's flying spot is a bit larger and captures the average of what it spans on the film, or electronically, for example scan line 1 as output is the mixture of lines 1 and 2 from the camera, scan line 2 is the mixture of lines 2 and 3, etc.

Vertical Interval Switching: Randomly switching from one video signal to another, will often result in a jump in the picture upon playback. The problem is compounded when the tape is copied. To avoid this problem, switching is best performed on synchronized signals during the vertical blanking retrace period, known also as the vertical interval. This allows complete replacement of one whole frame by a second whole frame resulting in a very smooth on-screen switch.

Vertical Interval Time Code: See VITC.

Vertical Resolution: The amount of detail that can be perceived in the vertical direction.

Vertical Retrace Interval: The time during which the electron beam is moved from the lower right corner to the upper left corner of the screen to draw the next field. In NTSC video, 1.4 milliseconds, or enough time to draw 21 scan lines, has been set aside for this purpose and to insert information to synchronize the electron beam with the transmitted picture for maintaining vertical hold. Typically the electron beam draws 240 odd lines from top to bottom, then 23 lines while it returns to the top, then draws 240 even lines, then 22 lines during retrace, and so on.

Vertical Rotation: Spot rotations according to time (i.e., a spot placed at 6:00AM on one day may rotate to 7:15AM on the next day, and 8:45AM the following day). The idea is to avoid having spots place at the same time every day.

Vertical Scan Rate: For non-interlaced video, this is the same as the frame rate. For interlaced video, it is usually one-half the field rate.

Vertical Squeeze Trick: Adjusting the vertical size control (height control) downward so all the scan lines (the raster) occupy a space of the desired aspect ratio (usually 16: 9). Used to play 16:9 enhanced DVD's on 4:3 TV sets with increased sharpness compared with the 4:3 setting on the

DVD player. Whether a particular TV set can use this technique is a matter of luck since the adjustment may be complicated or may have side effects such as misconvergence.

Vertical Sync: Synchronizing pulses used to define the end of one television field and the start of the next, occurring at a rate of approximately 59.94 Hz. This is the portion of the video signal that tells the decoder where the top of the picture is.

Vestigial Sideband: A method of encoding digital data onto a carrier for RF transmission. 8-VSB is used for over-the-air broadcasting of ATSC HDTV in the USA.

VfW: See Video for Windows.

VG: An IMDb notation to indicate that a particular title is a video game.

VGA (Video Graphic Adapter): An analog computer video signal format or equipment to produce or display same, using 480 visible scan lines each normally representing 640 pixels. The significance of this format is that the video signal is made up of the same total number of scan lines (525) transmitted at the same rate (scan rate) as NTSC video converted to a progressive scan format. If not confined to a broadcast channel, an (interlaced) NTSC video signal can also hold the detail of 640 or more pixels across. VGA signals cannot be sent directly into a standard NTSC video input.

VHF (Very High Frequency): Commonly, television channels 2 through 13.

VHS (Video Home System): The most popular system of videotape for home use. Developed by JCV.

VHS Hi-Fi: An improved stereo audio recording/playback system found on some camcorders and VCRs. Because the audio tracks are mixed and recorded with the video signal, audio only dubbing of these tracks is not possible.

VHS-C (VHS-Compact): A miniature version of the VHS tape format utilizing smaller cassettes that may also be played on standard VHS machines by using an adapter cartridge.

Video 8: Videocassette format that uses 8mm tape in a cassette. Also called Hi8.

Video Assist: The process of simultaneously recording filmed picture onto videotape by means of the same lens system in order to immediately evaluate a take as soon as it is completed.

Video Bandwidth: The range between the lowest and highest signal frequency of a given video signal. In general, the higher the video bandwidth, the better the quality of the picture. Video bandwidths used in studio work typically vary between 3 and 12 MHz. Consumer VCRs are generally capable of 3-5.5 MHz.

Video Camera: A camera which contains an electronic image sensor rather than photographic film. The lens focuses an image on an electronic tube or CCD chip. A camera has electronic circuitry which generates color and sync pulses. Most portable consumer cameras are equipped with a full complement of audio circuitry, e.g., microphone, audio amplifier and additional audio electronics. In order to obtain better quality images, a professional camera has three tubes or a triple CCD system, one for each basic color. Most professional cameras have a genlock input, which allows the camera to be synchronized to an external source. Some cameras also include basic character generators for titling purposes.

Video Carrier: A specific frequency that is modulated with video data before being mixed with the audio data and transmitted.

Video CD: An extension of CD based on MPEG-1 video and audio. Allows playback of near-VHS-quality video on a Video CD player, CD-i player, or computer with MPEG decoding capability.

Video Digital Effects: Electronic special effects that make it possible to alter video in a variety of ways, such as compressing, flipping and reversing polarity.

Video Editing: A procedure for combining selected portions of video footage in order to create a new, combined version. A variety of editing consoles are available. During video editing, special effects such as wipes, dissolves, inserts, etc. can be added. Professional editing is done using time code recorded on every frame of the magnetic tape allowing single frame accuracy. Audio editing is often carried out simultaneously with video editing.

Video Engineer: An individual who operates the video controls of a CCU to attain the best video quality.

Video Enhancer: A general term used to describe a device used to correct video image problems.

Video for Windows (VfW): The system software additions used for motion video playback in Microsoft Windows. Replaced in newer versions of Windows by DirectShow (formerly called ActiveMovie).

Video Gain Switch: A selector switch on a camera or CCU which increases gain or amplitude of a video signal—generally at the expense of some video quality. Used to compensate for the lack of light.

Video Gain: Amount of amplification for video signals, expressed in decibels (dB).

Video Head: The small signal-to-tape transfer device responsible for recording the video signal in a VTR. Video heads are mounted on a head wheel that rotates at a high rate of speed in relation to the videotape.

Video Interface Port: A digital video interface designed to simplify interfacing video ICs together. One portion is a digital video interface (based on BT.656) designed to simplify interfacing video ICs together. A second portion is a host processor interface. VIP is a VESA specification.

Video Leader: A visual and audio countdown at the

beginning of a videotape. Generally starts 10 seconds before the beginning of the show and visibly counts down to two seconds. Also called Countdown Leader.

Video Level: The strength or amplitude of a video signal.

Video Manager (VMG): In DVD-Video, the information and data to control one or more Video Title Sets (VTS) and Video Manager Menu (VMGM). It is composed of the Video Manager Information (VMGI), the Video Object Set for Video Manager Menu (VMGM_VOBS), and a backup of the VMGI (VMGI_BUP).

Video Mixer: A device used to combine video signals from two or more sources. Inputs are synchronized, then mixed along with various special effects patterns and shapes. A video mixer usually generates sync signals allowing genlocking of additional video sources to the first source.

Video Modulation: Converting a baseband video signal to an RF signal.

Video Module Interface: A digital video interface designed to simplify interfacing video ICs together. It is being replaced by VIP.

Video Monitor: A high-quality television display device, generally without a tuner or audio circuitry.

Video News Release (VNR): Essentially a press release packaged as a 2-3 broadcast news report and distributed (via satellite or cassette) to news organizations. The typical VNR will also include sound bites and b-roll separate from the edited piece that a news television station can edit themselves.

Video Noise: See Noise.

Video on Demand (VOD): Wide bandwidth cable or phone line allowing views to see or download full motion video in real time.

Video Printer: A special device used to capture a single frame of video to create a hard copy print.

Video Program System: VPS is used in some countries instead of PDC to control VCRs.

Video Projector: A display device which projects a video or computer image onto a large screen. The classic video projector has three primary color video tubes which converge on-screen to create the full color image. Single tube projectors eliminate convergence problems but compared to three tube systems, project a relatively lower quality image.

Video Prompter: A mechanical device that projects and advances text on a mirror directly in front of a camera lens, allowing talent to read lines while appearing to maintain eye contact with viewers.

Video Switcher: See Switcher.

Video Tap: When a video camera is mounted on the viewfinder of a motion picture camera, we then call that video camera a "video tap" It is used to allow others to view what the camera operator sees through the optical viewfinder. The image generated from the video tap is often combined with other recorded or generated images to preview a possible composite that is to be performed in post production.

Video Title Set (VTS): In DVD-Video, a collection of Titles and Video Title Set Menu (VTSM) to control 1 to 99 Titles. It is composed of the Video Title Set Information (VTSI), the Video Object Set for the Menu (VTSM_VOBS), the Video Object Set for the Title (VTST_VOBS), and a backup of the VTSI (VTSI_BUP).

Video Toaster: Software/hardware developed by NewTek for the Amiga Computer which produces affordable special effects.

Video Wall: A large array of several monitors, placed close to one another in the shape of a video screen or "wall." Each monitor is fed only part of the original video image by using a video-wall generating unit. This device is a digitally-based processor which converts the original analog video signal to digital, rescans, resamples and generates several individual analog video outputs for driving each array monitor separately. When viewed from a distance, the effect can be very dramatic.

VIDEO_TS: UDF file name used for video directory on disc volume. Files under this directory name contain pointers to the sectors on the disc which hold the program streams.

Video-Black-Video: See VBV.

Videocassette: A container holding both the feed and take-up reels which is inserted into a videotape recorder and threaded automatically.

Videographer: The person overseeing all aspects of the video image—including lighting and the operation of the camera.

Videography: The characteristics of the video camera.

Videophile: Someone with an avid interest in watching videos or in making video recordings. Videophiles are often very particular about audio quality, picture quality, and aspect ratio to the point of snobbishness.

Videospace: All visual elements in a production which interact to create the experience of reality.

Vidicon: A one-popular type of camera tube known for its reliability and long life.

Viewers Per Viewing Household (VPVH): The number of viewing persons per tuning household. This figure is usually reported as "per 1,000 viewing households" or "per 100 viewing households." Sometimes referred to as VPH.

Viewfinder: A video monitor attached directly to a video camera.

Vignette: Visual special effect whereby viewers see images through a perceived keyhole, heart shape, diamond: whatever. In low-budget form, achieved by aiming camera through cutout of desired vignette.

VIP: See Video Interface Port.

Virgin Stock: Magnetic tape onto which nothing has yet been recorded.

Virtual Set: A virtual set is one that is an alternative to a set built with nails, wood, plaster or other construction materials. Virtual sets are generated electronically and can be manipulated by means of 3-D image processing to simulate what would otherwise be a real physical set.

Visual Effects: Alterations to a film's images during post-production.

VITC (Vertical Interval Time Code): A popular method for recording time code onto videotape. A time code address for each video frame is inserted in the vertical interval (the vertical blanking retrace period) of the video signal, where it is invisible on-screen yet easily retrieved, even when a helical scanning VCR is in pause mode. The most common form of VITC is SMPTE-VITC.

VLC (Variable Length Coding): See Huffman Coding.

VL-mount: Standardized lens mount for camcorders allowing the use of different types of lenses, all of which can successfully interface with the camera electronics.

VMGI (Video Manager Information): Information required to manage one or more Video Title Sets and Video Manager Menu areas. This is non real time data located at the start of the Video Manager area.

VMI: See Video Module Interface.

VNR: See Video News Release.

VOB (Video Object): A single, complete file composed of multiplexed Video, Audio, Sub-picture, PCI and DSI elementary streams, and consisting of an integer number of VOBUs.

VOBS (Video Object Set): A collection of one or more VOBs.

VOBU (Video Object Unit): A small (between 0.4 and 1.0 seconds) physical unit of DVD-Video data storage, usually the length of one GOP, that begins with a Navigation pack (NV_PCK) and usually includes an integer number of GOPs.

Voice-over: Narration or non-synchronous dialog taking place over the action onscreen.

Volt: The standard unit for measuring the difference of potential between two points in an electronic circuit.

Volume Management Information: Identifies disc side and content type.

Volume Space: Collection of sectors that make the volume. Not all sectors on the disc comprise the volume. Some near the inner and out spiral are used as leader.

Volume: How loud a sound is.

Volumetric: Effects that are drawn within a 3-D volume: not flat plane in space. For example, Volumetric Fog fills an area, and can be moved inside and through, like a real fog bank.

VPS: See Video Program System.

VPVH: See Viewers Per Viewing Household.

VSB (Vestigial side band): VSB is an analog modulation technique used to reduce the amount of spectrum needed to transmit information through cable TV, or over-the-air broadcasts used in the NTSC (analog) standard.

VSDA: Video Software Dealers Association.

VSM: See Velocity Scan Modulation.

VTR: Videotape recorder.

VTSI (Video Title Set Information): Information required to manage one or more Titles and Video Title Set Menus. This is non real time data located at the start of the Video Title Set.

VU (Volume Unit): Unit of measure for audio level or signal strength.

VU Meter: An instrument (meter) that measures the loudness of sound in terms of decibels and percentage of modulation.

VVV (Video-Video-Video): The standard preview mode used in conventional editing.

W

WAD: A WAD file is a file containing a group of textures. Half-Life WADs contain textures that all have individual 8-bit (256 color) palettes. Also referred to as a WAD3.

WAEA: See World Airline Entertainment Association.

Walk-through: A rough rehearsal in which no cameras are used. Generally the actors walk from place to place and check their actions without speaking any dialogue.

Walla: Background ambience or noises added to create the illusion of sound taking place outside of the main action in a picture. Historically, when a script called for "crowd unrest" or "murmuring," the extras would be required to mumble the word "rhubarb," as this produced the required effect.

Wardrobe Department: Section of a production's crew concerned with costumes: costume designer, costumer, and costume supervisor.

Wardrobe Person: Individual responsible for selecting and supplying clothes and accessories for talent.

Warm: A color picture that contains a noticeable amount of red or yellow.

Waste: Shining a light on an object, then slowing turning the light away so that some of the light will miss or fall off the object.

Watermark: Information hidden as "invisible noise" or "inaudible noise" in a video or audio signal.

Watermarking: A system of invisibly inserting information into a digital media file that identifies the file's copyright information.

Watt: Unit of electrical power equal to one volt across a resistance of one ohm, or one volt at a current of one ampere.

WAV: An uncompressed audio file format.

Wave: A regular variation in electrical signal level or sound pressure level.

Waveform Monitor: Specialized oscilloscope testing device providing a graphic display of a video signal's strength. Plus, like a sophisticated light meter, aids in precise setting of picture's maximum brightness level for optimum contrast.

Weave: Method of de-interlacing where an intervening scan line is taken from the next field.

WebTV: WebTV Networks, Inc. manufactures set-top boxes that allow users/viewers to access the Internet on their NTSC TV receivers.

Wedge Mount: Mounting device enabling cameras to be rapidly mounted and dismounted from cameras pedestals or tripods.

Wedges: Wood wedges cut from 2x4 lumber which is used for leveling and stabilizing.

Wet Gate: A contact printing method, made on a specially equipped printing machine, where the film is in a liquid that temporarily fills in any scratches on the base, preventing them from refracting light and showing up in the print. Commonly, answer prints are printed with a wet gate. Labs often charge a little extra for wet gate printing.

WGA: See Writer's Guild of America.

Whip Pan: An extremely fast pan, incorporating much motion blur. The term refers to the "whipping" action that the camera operator uses to move the camera.

Whip: A section of feeder cable siamesed off the main line to a secondary location.

White Balance Adjustment: In the light of a particular color temperature, to adjust the white levels of the R, G, and B channels of a color video camera so that any white object shot in that light is reproduced as a truly white image. See also Color Temperature.

White Book: The document from Sony, Philips, and JVC, begun in 1993 that extended the Red Book compact disc format to include digital video in MPEG-1 format. Commonly called a Video CD.

White Flag: The means whereby a (12") CAV laser disk is encoded so that the even and odd fields match when the player displays a still frame. Without white flagging, the laser player may do a still frame with the odd interlaced field taken from what was one film frame and the even interlaced field taken from the next film frame. The result is a flickering double exposure effect.

White Level: This level defines what white is for the particular video system.

White Noise: A signal having an equal amount of energy per hertz.

White Shading: When shooting a white object, the upper and lower portions of the screen may appear magenta or green while the central portion appears white, depending on the performance of the camera lens. This is called white shading.

Wide Angle: Any lens with a focal length significantly less than the standard length for the format in which it is used. For 35mm still cameras and one inch television cameras, the standard length is 50mm.

Wide Lens: A lens with a focal length smaller than 25mm in 16mm, or 50mm in 35mm, which, like looking into the wrong end of a pair of binoculars, provides an extended view of a large area.

Wide Shot (WS): A picture showing a subject in the context of the surroundings to establish the relationship between the subject and the surroundings.

Wide-angle Converter: Lens attachment that gives a wider angle of view by decreasing the lens's focal length.

Wide-angle Lens: A lens which gives a wide field of view and exaggerates depth. The lens tends to exaggerate perspective, making straight lines towards the edges of the screen bulge outward.

Widescreen: Refers to a video program whose picture has a wider aspect ratio than 4:3.

Wild Line: A line of dialogue, recorded either on set or at a looping stage, without any picture running.

Wild Sound: Sound recorded after the visuals and edited into the master to enhance realism.

Wind Screen: A thin soft foam cover for microphones which reduces the noise made by wind striking the microphone.

Window Dub: A copy of the original camera tape with time code numbers visually displayed. Also called a "burn in."

Window: A usually rectangular section within an entire screen or picture.

Window: The length of time, usually expressed in weeks, in which a syndicated movie or special will air. The national rating is not available until after the final week of a window.

Windowbox: Refers to a video program where the picture is zoomed out so far that unused screen area appears on all four sides. The purported reason for doing this is to ensure that the entire picture can be seen even on TV sets with a lot of overscan.

Wipe: A transition from one shot to another in which the second shot peels off the first one, an effect comparable to that of a windshield wiper. Wipes are varied in form, ranging from a single straight edge moving vertically, horizontally or diagonally to multiple edges and geometric patterns moving simultaneously in different sectors of the frame. Rarely used in contemporary film, but common in films from the 1930s and 1940s.

Wire Frame: A stage in computer-generated imagery wherein the surface of objects is represented with polygonal lines (wires). The wire frame will be covered with surfaces when the animation is rendered.

Wireless Microphone: A microphone that has a built-in, low-power transmitter, or that is connected to a transmitter. Wireless microphones are commonly used when a microphone cord would create a problem.

Woofer: Large speaker, effective in reproducing bass.

Working Title: The name by which a movie is known while it is being made. This is sometimes different from the title with which it is released.

Workprint: A duplicate of film footage, used during the editing process in order to preserve the original intact until the final cut.

Workstation: This term generally refers to a disk-based audio recording and editing system.

World Airline Entertainment Association (WAEA): Discs produced for use in airplanes contain extra information in a WAEA directory. The in-flight entertainment working group of the WAEA petitioned the DVD Forum to assign region 8 to discs intended for in-flight use.

World System Teletext (WST): BT.653 525-line and 625-line system B teletext.

WORM (Write Once Read Many [Times]): A videodisc system that permanently writes information onto a disk.

WOW and Flutter: Sound distortions consisting of a slow rise and fall of pitch, caused by speed variations in audio/video playback system.

Wrangler: Individual who is responsible for the care and control of entities used on a set that can't be spoken with.

Wrap: The span of the tape path along which the tape and head are in contact. More often, this refers to securing equipment at the end of the day or when work is completed at a particular set or location. Also to finish shooting, either for the day or the entire production.

Writer's Guild of America (WGA): The Writers Guild of America is the sole collective bargaining representative for writers in the motion picture, broadcast, cable, interactive and new media industries. It has numerous affiliation agreements with other U.S. and international writing organizations and is in the forefront of the debates concerning economic and creative rights for writers.

Writing Period: The time during which a writer is to complete his work. During this time the writer's services are generally exclusive to the production that has hired him.

Written By: The credit given when one or several writers have created both the story and the screenplay, and there is no source material. The credit is also given in television if the writer has created both the story and the teleplay.

WST: See World System Teletext.

WYSIWYG: An acronym (pronounced "wizzywig") for what-you-see-is-what-you-get. Refers to a computer screen representation of text or graphics being the same as the subsequent hardcopy printout.

X

X-axis: The horizontal line (axis) of a graph or chart.

XCU: See Extreme Close Up.

Xenon: A very bright, daylight balanced projection lamp, or a projector with a xenon lamp. A xenon lamp is not interchangeable with a tungsten lamp or arch lamp, but requires a different lamp housing on the projector. Because xenon lamps are daylight balanced it is sometimes advisable with color film to have the lab make a print that is balanced for xenon. This is sometimes called a 5,400K print, the color temperature of daylight.

XGA: An analog computer video format with 768 visible scan lines each normally representing 1024 pixels across.

XLR: Three-pin plug for three-conductor "balanced" audio cable, employed with high-quality microphones, mixers, and other audio equipment. Also called Cannon.

XLS: See Extreme Long shot.

XVCD: A non-standard variation of VCD.

XXX: An informal voluntary certificate for a pornographic film, indicating large amounts of explicit sex.

X-Y Microphone: A stereo microphone technique using two directional microphones aimed in a Y-pattern.

X-Y Pickup: A microphone configuration used in stereo recording in which two directional microphones are mounted next to each other one directed 45 degrees to the right and one 45 degrees to the left.

Y

Y M C Numbers: The "timing lights" that the negative is printed at, representing the numerical value of the red, blue and green lights used in the printing process. These printing lights are referred to by their complements: yellow, magenta, and cyan.

Y, C_r, C_b: The digital luminance and color difference signals in ITU-R BT.601 coding. The Y luminance signal is sampled at 13.5 MHz and the two color difference signals are sampled at 6.75 MHz co-sited with one of the luminance samples. C_r is the digitized version of the analogue component R-Y, C_b the digitized version of B-Y.

Y, R-Y, B-Y: The luminance, Y and color difference signals, R-Y and B-Y, of analog component television. Y is pure luminance information while the two color difference signals together provide the color information.

Y/C (Luminance and Chrominance): A term used to describe the separation of video signal components used in systems such as Hi-8 and S-VHS. Generically called S-Video.

Y/C/S: Method of transmitting color video over two conductors, one carrying luminance (Y) and the other carrying color (C) information.

Y/C: A video signal in which the brightness (luma, Y) and color (chroma, C) signals are separated. Also called S-video.

Y: Symbol for luminance, or brightness, portion of a video signal; the complete color video signal consists of R, G, B, and Y.

Y'IQ, YIQ: Y'IQ is a color space optionally used by the NTSC video system. The Y' component is the black-and-white portion of the image. The I and Q parts are the color difference components; these are effectively nothing more than color placed over the black and white, or luma, component. Many people use the YIQ notation rather than Y'IQ or Y'I'Q.' The technically correct notation is Y'I'Q' since all three components are derived from R'G'B.'

Y-adapter: An audio connector which combines two signals or splits one signal into two.

Yagi: A specific design for an outdoor TV antenna.

Yarn: Slang for an apocryphal story.

Y-axis: The vertical line on a graph or chart, usually containing the values for data points.

Yellow Book: The document produced in 1985 by Sony and Philips that extended the Red Book compact disc format to include digital data for use by a computer. Commonly called a CD-ROM.

Yo-yo Shots: Repeated zooming in and out on subjects. An annoying effect associated with amateur productions.

Y, $P_{b,}$ P_r: A component analog video signal containing one luma and two chroma components. Often referred to loosely as YUV or Y, B-Y, R-Y.

YUV12: Intel's notation for MPEG-1 4**:** 2**:** 0 YCbCr stored in memory in a planar format. The picture is divided into blocks, with each block comprising 2 x 2 samples. For each block, four 8-bit values of Y, one 8-bit value of Cb, and one 8-bit value of Cr are assigned. The result is an average of 12 bits per pixel.

YUV9: Intel's 4**:** 1**:** 0 YCbCr format. The picture is divided into blocks, with each block comprising 4 x 4 samples. For each block, sixteen 8-bit values of Y, one 8-bit value of Cb, and one 8-bit value of Cr are assigned. The result is an average of 9 bits per pixel.

YUY2: Intel's notation for 4**:** 2**:** 2 YCbCr format.

Z

Zapping: Avoiding seeing a television commercial by channel switching, usually by remote control. Also refers to the deletion of commercials during VCR recording.

Z-axis: The front-to-back dimension a data chart. Usually contains values for data points.

Zebra Pattern: In a video camera, striped patterns which appear in the viewfinder screen to indicate areas of the image where the video level is higher than a certain value.

Zero Cut: A method of negative cutting specifically for blow up, where every shot is given Frame Handles so that the registration pin of the printer is never engaging with a splice, which can cause the image to wobble at the cut. It is most commonly used when you are blowing up from 16mm to 35mm. Zero cut should be done only if really necessary, because the lab can only print the film as an optical, which is far more expensive than a contact print. Zero cutting is a little more complicated than standard A&B rolls, so the negative cutter also charges more for it.

Zero Cut: A method of preparing A and B rolls for printing in which the original shots overlap several frames or more. The change from one roll to another to match the edited workprint is done automatically by the printer.

Zip Pan: See Swish Pan.

Zipper: See Creepy-Crawlies.

Zoom Lens: A camera lens of variable focal length that can be changed while shooting, thus allowing the camera to photograph from wide-angle to telephoto range in the same shot.

Zoom Ratio: The ratio of the longest focal length to the shortest focal length of a zoom lens.

Zoom Shot: A shot with a zoom lens, which can make a

film image appear closer (zoom in) or farther away (zoom out) by varying the focal length of the lens.

Zoom: Shot in which the magnification of the objects by the camera's lenses is increased (zoom in) or decreased (zoom out/back).

Zoomed Video Port: Used on laptops, the ZV Port is a point-to-point unidirectional bus between the PC Card host adapter and the graphics controller, enabling video data to be transferred real-time directly from the PC Card into the graphics frame buffer. The PC Card host adapter has a special multimedia mode configuration. If a non-ZV PC Card is plugged into the slot, the host adapter is not switched into the multimedia mode, and the PC Card behaves as expected. Once a ZV card has been plugged in and the host adapter has been switched to the multimedia mode, the pin assignments change.

Zweiton: A technique of implementing stereo or dual-mono audio for NTSC and PAL video. One FM subcarrier transmits a L+R signal, and a second FM subcarrier transmits a R signal (for stereo) or a second L+R signal. It is discussed in BS.707, and is similar to the BTSC technique.

Contributing Organizations and Companies

Half-Life; The New School; Internet Movie Database, Today's Video; Adobe Systems; Epic Multimedia; KCTS; Eastman Kodak; Annenberg/CPB; Quantel, Inc; FilmSecrets; Special Interest Video Sales Group; TV Cameraman; Video Demystified; Videofonics; Zerocut; University of Maryland; DVD Made Easy; CyberCollege; Encoda Systems; Video Essentials. EZ DVD Advisor; iLearn; Compression Works; TV Guide; Sweetwater; Television: Critical Methods and Applications, University of Alabama; Calvert; Net Composite; Grantastic Designs; Third Millennium Entertainment; SVN, Inc.; Silcone Valley North, Inc.; Motto; Virtual Worlds; Apple; Cleaver Brooks; EKEDA; Prolink USA; System 16 Arcade Museum; Webopedia; California Department of Education; Eye on Video; Tippett Studio; Florida Department of Education; Internet Campus; Planet Quake; Intellisoft; Australian Centre for the Moving Image; Sympatico; The Scratch Post; Introduction to Film, Berkeley; Arts Act; Animation USA; Demon Internet; Utah State University; DUMC Department of Surgery Digital Media Guide; Kitezh; Bergen; Profim, S.A.; Meisler; Aigcorp; Hewlett Packard; Digital Exposure; Digital Director; Mississippi State University; 4-K Associates; Cobalt Systems; Gallery Direct; Chem Co.; The Way Computer Graphics Works; SalmonIdaho; Digital TV Zone; Design In You; High End 3D; Canon; Rendering with Radiance; 3D Technology; CyberEdge; SoftImage Effect Web; ACM SIGGRAPH; DVD Made Easy; Nonoctave; Source Forge; University of Toronto; Audio Education; Video Guys; Products, Services, and Ideas.com; Realsoft; IT-Director; Alias/Wavefront; IPSI Imaging; The Ohio State University; Wallstreet Webs; Computer Science Society; Survey of Medical VR; Soft Landmark; OpenGL Performer Getting Started Guide; Ripon Printers; Future Technology; University of Maryland, Baltimore County; Unicom Telecommunication; DIT Imaging; Guru 3D; Bib2Web; PLASMA; Crossbow Technology; JPG Digital; The APECS Performance Animation System; Directron; International Hydrocut; Domain Decomposition of Geometrical Objects; Vintage Ink & Paint; Media Value; Internet Campus; Alaska Fisheries Information Network; University of California Los Angeles; Botics Computer Consulting.

Bibliography
Selected Books

Adobe Systems, Inc. Adobe Photoshop 5.5: Classroom in a Book. Adobe Press. 1999.

Finch, Christopher. Special Effects: Creating Movie Magic. Cross River Press. New York 1984.

Fry, Ron, and Pamela Rourzon. The Saga of Special Effects. Prentice-Hall International, New Jersey. 1977.

Gardner, Garth. Informal Computer-Art Education. Unpublished Dissertation. Ohio State University. 1995.

Gardner, Garth. Careers in Computer Graphics and Animation. GGC Publishing. Washington, DC. 2000.

Gardner, Garth. Computer Graphics and Animation: History, Careers, Expert Advice. GGC Publishing, Washington, DC. 2002

Kerlow, Isaac Victor. The Art Of 3-D Computer Animation and Imaging, John Wiley & Sons, Inc. New York. 2002.

Kerlow, Isaac Victor, and Judson Rosebush. Computer Graphics for Designers and Artists. Van Nostrand Reinhold. New York. 1986.

Laybourne, Kit. The Animation Book. Crown Traders Paperbacks, New York. 1979.

Lopuck, Lisa. Designing Multimedia. Peachpit Press, Berkeley, CA. 1996.

Lord, Peter, and Brian Sibley. Creating 3-D Animation: The Aardman Book of Filmmaking. Harry N. Abrams, Inc. New York. 1998.

Masson, Terrence. CG 101: A Computer Graphic Industry Reference. New Riders Publishing, Indianapolis. 1999.

Morrison, Mike. Becoming A Computer Animator. Sams Publishing, Indianapolis. 1994.

Olson, Robert. Art Direction For Film and Video. Focal Press, Boston London. 1993.

Roberts, Jason, and Phil Gross. Director Demystified 7. Macromedia Press, Peachpit Press, Berkeley, CA. 1999.

Webber, Marilyn. Gardner's Guide to Animation Scriptwriting: The Writer's Road Map. GGC Publishing, Washington D.C. 2000.

Webber, Marilyn. Gardner's Guide to Feature Animation Screenwriting: The Writer's Road Map. GGC Publishing, Washington D.C. 2002.

Webber, Marilyn. Gardner's Guide to Screenwriting: The Writer's Road Map. GGC Publishing, Washington D.C. 2000.

Webber, Marilyn. Gardner's Guide to Television Scriptwriting: The Writer's Road Map. GGC Publishing, Washington D.C. 2002.

Zettl, Herbert. Television Production Handbook, 6th Edition. Wadsworth Publishing Company. Belmont, CA. 1997.

GARTH GARDNER COMPANY

GGC publishing

Washington DC, USA · London, UK

Other GARDNER'S Guides

Gardner's Guide to Careers in Computer Graphics & Animation

Gardner's Guide to Computer Graphics and Animation: History, Careers, Expert Advice

Gardner's Guide to Television Scriptwriting: The Writer's Road Map

Gardner's Guide to Animation Scriptwriting: The Writer's Road Map

Gardner's Guide to Screenwriting: The Writer's Road Map

Gardner's Guide to Feature Animation Writing: The Writer's Road Map

Gardner's Guide to Colleges for Multimedia & Animation

Gardner's Storyboard Sketchbook

Gardner's Guide to Multimedia and Animation Studios

Gardner's Guide to internships in New Media

orders@ggcinc.com
www.gogardner.com